U.S. MARINES IN LEBANON 1982-1984

by

Benis M. Frank

HISTORY AND MUSEUMS DIVISION
HEADQUARTERS, U.S. MARINE CORPS
WASHINGTON, D.C.

1987

Library of Congress Card No. 87-619851
USMC PCN 190 003098 00

For sale by the Superintendent of Documents, U.S. Government Printing Office
Washington, D.C. 20402

Foreword

This book is a straightforward account of the deployment of Marines to Lebanon in the period 1982-1984. The story begins with the landing of the 32d Marine Amphibious Unit (32d MAU) in Beirut in August 1982 at the request of the Lebanese Government to assist, together with French and Italian military units, in supervising the evacuation of the Palestine Liberation Organization. It ends in February 1984 with the withdrawal of the 22d Marine Amphibious Unit following the effective end of its mission and the nearly complete breakdown of order in Lebanon. In between is an ambiguous Marine mission of presence of 18 months' duration. Together with the British, French, and Italian members of the Multi-National Force, the Marines attempted, as "peacekeepers," to assist the Lebanese Government in achieving stability and ending the factional fighting which has all but destroyed Lebanon as a viable political entity.

For any number of reasons, none of which are the concern of this book, the mission of peacekeeping failed, and in the process, those who were there to help Lebanon achieve the peace so many Lebanese wanted—but too many others did not—were sorely tried and severely mauled. As a history strictly of the Marines' role in Lebanon, this book does not deal with the major, high-level decisions of the administration which put and kept Marines in that country. Nor does the book deal with American diplomatic efforts in the Middle East in this period except in those instances when the MAU Marines were directly involved. This is simply the story of Marine Corps presence and operations in Lebanon for the period concerned. It draws no conclusions.

The author, Benis M. Frank, is the head of the Marine Corps Oral History Program. As such, in two trips to Beirut and three to Camp Lejeune, he interviewed the major members of the staffs and commands of the three MAUs (22d, 24th, and 32d) which were deployed to Beirut. For his second trip to Beirut in October 1983, he went by way of Grenada, where he joined the 22d MAU in transiting the Atlantic, conducting interviews about the Grenada operation en route to Lebanon. Mr. Frank graduated from the University of Connecticut in 1949 with a bachelor of arts degree in history. His schooling was interrupted by World War II, in which he served as an enlisted Marine with the 1st Marine Division in the Peleliu and Okinawa operations and the occupation of North China. He was a candidate for a master of arts degree in international relations at Clark University when he left school in 1950 to return to active service in the Korean War as a commissioned officer, again serving with the 1st Marine Division.

Mr. Frank joined the Marine Corps Historical Program in 1961. He is the coauthor of *Victory and Occupation*, the final volume of the official five-volume series, *History of U.S. Marine Corps Operations in World War II*, and author of *Halsey*; *Okinawa, Touchstone to Victory*; *Okinawa: The Great Island Battle*; and *Denig's Demons and How They*

Grew: A History of Marine Corps Combat Correspondents, Photographers, and Artists. He pioneered the Marine Corps Oral History Program and has headed it since 1966.

In the interests of accuracy and objectivity, the History and Museums Division welcomes comments on this history from interested individuals.

Edwin H. Simmons
Brigadier General, U.S. Marine Corps (Ret.)
Director of Marine Corps History and Museums

Preface

U.S. Marines in Lebanon, 1982-1984, is based primarily on the monthly command chronologies and biweekly situation reports of the Marine amphibious units which were deployed to Lebanon as well as other related official documentation, all of which resides in the archives of the Marine Corps Historical Center. Although none of the information in this history is classified, some of the documentation on which it is based remains so. A considerable number of "issue-oriented" oral history interviews concerned with the deployments were also used in the preparation of this book.

Following the return of the 32d MAU from Lebanon and before its redeployment in early 1983, the author began a series of interviews with the key personnel in all the MAUs deployed to Lebanon to augment the paper record of this 18-month period in Marine Corps history. Before *U.S. Marines in Lebanon* was completed, a total of 119 interviews had been conducted. They are now accessioned in the Marine Corps Oral History Collection.

The author is grateful to a number of individuals for their professional, administrative, and moral support during the research and writing phases for this book. First, Mrs. Alexandra B. Chaker, his assistant in the Oral History Section, prepared the initial manuscript for typography and was in all other ways entirely supportive. Mrs. Ann A. Ferrante, of the Reference Section, responded nobly when called upon to research the voluminous Lebanon files in the section. Similarly, Mrs. Joyce Bonnett, the Center archivist, consistently provided pertinent documentation as soon as it arrived in the Center, as did Miss Evelyn A. Englander, head librarian.

The various production phases this volume went through before publication were professionally handled by the head of the Publications Production Section, Mr. Robert E. Struder. His able associate, Mrs. Catherine A. Kerns, set the manuscript in type. Mr. William S. Hill, the History and Museums Division graphics specialist, is responsible for the design and layout of this book. The author prepared the index.

The author gratefully acknowledges the support of the Director of Marine Corps History and Museums, Brigadier General Edwin H. Simmons, who conceived of this project and kept the author's "feet to the fire" to ensure completion of a publishable, factual, and objective manuscript. Two Deputies for Marine Corps History, Colonels John G. Miller and James R. Williams, also read, commented on, and shepherded the project to its completion. Gratitude is also extended to Mr. Henry I. Shaw, Jr., Chief Historian, who was the author's mentor and coauthor many years ago in writing *Victory and Occupation*, and who unfailingly and continually offered his considerable expertise in Marine Corps history, research and writing, and his extensive editorial guidance.

The author also extends his appreciation to the former commanders of the 22d, 24th, and 32d Marine Amphibious Units—Brigadier Generals Jim R. Joy and James M. Mead, and Colonels Thomas M. Stokes, Jr., and Timothy J. Geraghty—for having reviewed and commented on the draft manuscript of this book and for their hospitality and cooperation when he visited their commands to conduct oral history interviews. Major Jack L. Farmer, Assistant S-3 in the 32d MAU and S-3 of the 22d MAU, read the draft, was interviewed several times for the history, and provided considerable background information to enhance the story, as did Commander George W. Pucciarelli, CHC, USN, 24th MAU chaplain at the time of the Beirut tragedy. The expert reviews and comments of both J. Robert Moskin, author of *The U.S. Marine Corps Story*, and Larry Pintak, former CBS Mid-East correspondent who covered Beirut during the Marine deployments there are noted with deep appreciation. Similarly, the author is grateful to Major Fred T. Lash, who headed the MAUs' Joint Public Affairs Bureau in Beirut, for having hosted and guided him when he was in Lebanon and for lending his collection of photographs and political cartoons for use in this book. Three members of the Department of Plans, Policies, and Operations at Headquarters, U.S. Marine Corps—Colonel Gerald J. Oberndorfer and Lieutenant Colonels Arthur S. Weber, Jr., and Robert P. Mauskapff—were also quite helpful in reviewing the draft manuscript. The author also acknowledges with gratitude the cooperation and insights given by all of those serving Marines who were interviewed about their Beirut experiences, and those who read and meaningfully commented on the draft. This is, in a large way, really their history.

It would be totally ungracious for the author not to acknowledge the considerable moral support he received from his wife Marylou, as he wrote this history. She read the draft manuscript and made cogent recommendations which were sage, pertinent, and gratefully accepted. The author, however, is responsible for the contents of this work and any errors of omission or commission which appear.

Finally, this book is dedicated to those United States Marines, sailors, and soldiers who gave "presence"—and their lives—in Beirut and are now no more.

BENIS M. FRANK

Table of Contents

Maps

Defense Mapping Servic

Beirut and its environs.

CHAPTER 1

The Bombing

Dawn broke over Beirut at 0524 local time on Sunday, 23 October 1983.[1] The temperature was already a comfortable 77 degrees F, but perhaps a bit warm for 24th Marine Amphibious Unit (MAU) sentries posted around the perimeter of the MAU headquarters compound at Beirut International Airport. They were in full combat gear—helmets, upper body armor—and carried individual weapons. Since it was Sunday, the compound was relatively quiet for a modified holiday routine was in effect. Reveille would not go until 0630, and brunch would be served between 0800 and 1000. In the afternoon, there would be time to write letters, read, and perhaps toss a football about. In the afternoon there might be a barbecue—hamburgers, hot dogs. and all the trimmings.[2]

Relatively little traffic was observed in the early morning hours on the airport road which runs between Beirut and the airport terminal. This road is just west of and runs parallel to the MAU compound. The Marines had been warned to be alert for suspicious looking vehicles which might, in fact, be terrorist car bombs. And so Lance Corporal Eddie A. DiFranco, manning Post 6 (See Figure 1), one of the two posts in front of and south of the building housing the headquarters compound and attached elements of BLT 1/8 (Battalion Landing Team 1/8, built around the 1st Battalion, 8th Marines), closely watched a yellow Mercedes Benz stake-bed truck, which entered the parking lot south of his post. The truck circled the lot once, then departed, turning south at the gate and heading towards the terminal.

A little less than an hour later—it went down in the reports as 0622—DiFranco saw what appeared to be the same truck enter the same parking lot. This time, the vehicle accelerated to the west, circled the lot once, then headed toward the wire barricade separating the parking lot from the BLT building. Turning right, it ran over the wire barricade and sped

A low oblique of the BLT headquarters building, before the October bombing.

Photo from the *Long Commission Report*

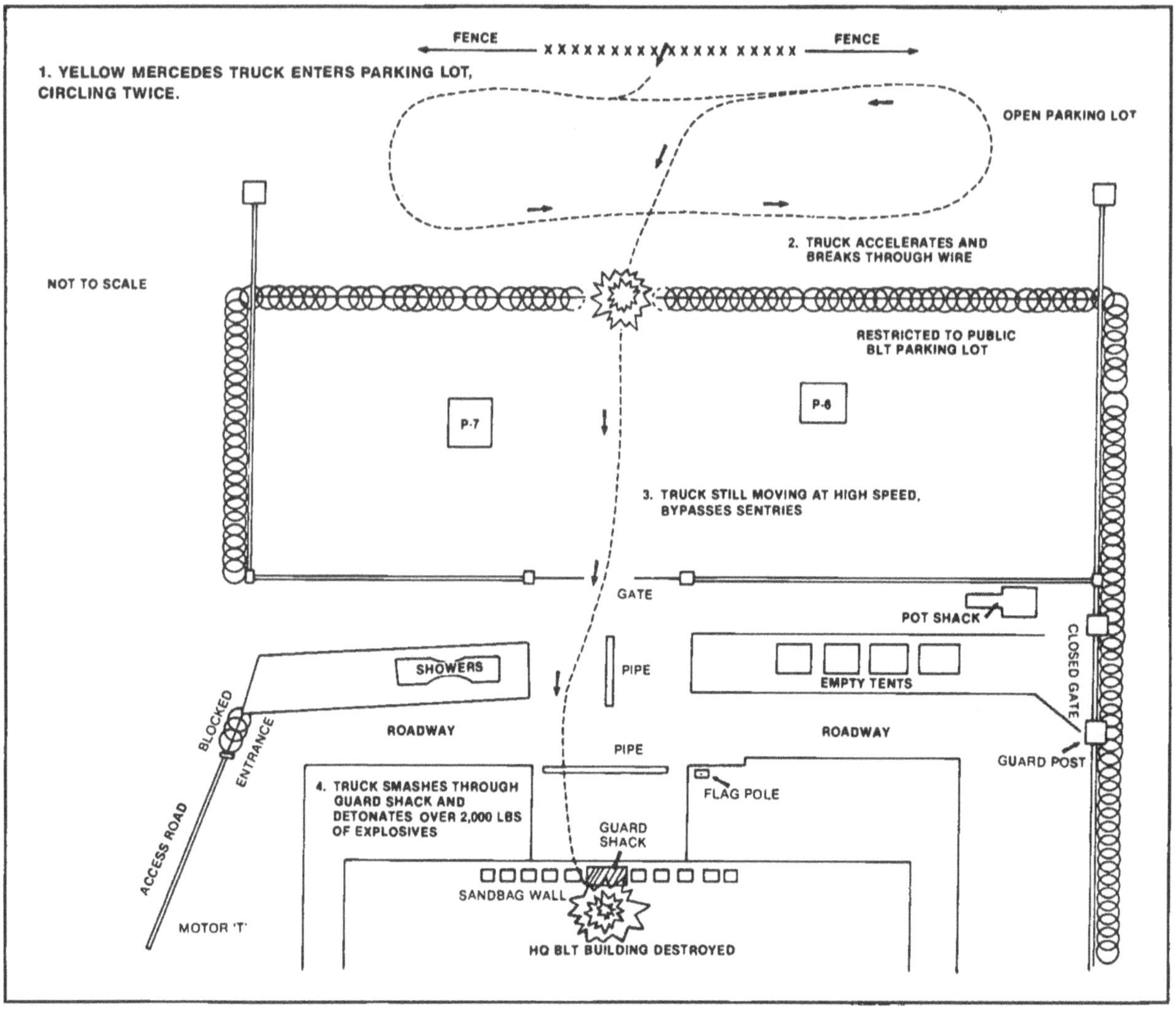

Figure 1 From *Long Commission Report*

Sketch map of the route taken by the terrorist bomber on the morning of 23 October 1983.

between Posts 6 and 7 into the lobby of the building, where it detonated with the explosive force of more than 12,000 pounds of TNT.

Manning Post 7 was Lance Corporal Henry P. Linkkila, who heard the truck as it sped across the concertina fence. He inserted a magazine into his M-16 rifle. He chambered a round and shouldered his weapon, but could not fire. The truck had already entered the building.

Lance Corporal John W. Berthiaume was guarding Post 5, at the fence just below the southwest corner of the BLT headquarters. He correctly guessed the truck's mission, but could not react in time either to fire at the truck or to take cover in his guard bunker. He was knocked to the ground by the explosion.

Sergeant of the Guard Stephen E. Russell was at the main entrance of the building at his post, a small sandbagged structure that looked toward the back entrance to the building, when he heard the truck as the driver revved up its engine for the dash into the lobby. Russell turned to see the vehicle pass through the permanent fence encircling the compound, and head straight for his post. He wondered what the truck was doing inside the compound. Almost as quickly, he recognized that it was a threat. He ran from his guard shack across the lobby toward the rear entrance, yelling, "Hit the deck! Hit the deck!" Glancing over his shoulder as he ran, he saw the truck smash through his guard shack. A second or two later the truck exploded, blowing him into the air and out of the building. Severely injured, Russell regained consciousness and found himself in the road outside the BLT headquarters with debris from the explosion all around him.

It had finally happened. An explosive-laden truck had been driven into the lobby of a building billet-

USMC Photo by SSgt Robert E. Kline

The front entrance of the BLT headquarters building in August 1983 when occupied by BLT 1/8. The truck bomb entered at the point where the jeep is shown parked.

ing more than 300 men, and detonated. The explosion had collapsed the BLT building, reducing it to rubble in seconds.

When the last body had been retrieved from the ruins and the final death count had been tallied, it reached a total of 241 Americans. Of this number, 220 were Marines; the remainder, Navy medical personnel and soldiers assigned to the MAU. For the Marines, this was the highest loss of life in a single day since D Day on Iwo Jima in 1945.

The suicide attack by a single terrorist changed the course of American presence in Lebanon.

The rear (north) side of the BLT building immediately after the bombing.

CHAPTER 2

Beirut I—Evacuating the PLO 25 August-10 September 1982

Marines had been in Beirut before—in 1958, to be exact.[1] In July of that year, Lebanon was threatened by a civil war between Christian Maronites and Muslims. Additionally, Lebanon faced a potential Syrian invasion in support of the Muslims. Accordingly, on 14 July, in response to the internal and external threats, Lebanese President Camille Chamoun requested American and British assistance. That same day, President Dwight D. Eisenhower consulted with the Joint Chiefs of Staff about the Lebanon crisis, studied their recommendations, and ordered the deployment of U.S. troops to Lebanon.

As a matter of happenstance as well as of contingency planning, three Marine battalion landing teams were then in the Mediterranean. Afloat just north of Malta was Lieutenant Colonel John H. Brickley's BLT 1/8, its Mediterranean deployment near an end, ready to be relieved by the recently arrived BLT 3/6, commanded by Lieutenant Colonel Robert M. Jenkins, on board ships steaming from Crete to Athens. The third BLT was 2/2, Lieutenant Colonel Harry A. Hadd commanding, off the coast of Crete and closest of the three to Lebanon. Another Marine command in the Mediterranean was Brigadier General Sidney S. Wade's 2d Provisional Marine Force, which had been formed from troops of the 2d Marine Division at Camp Lejeune in January for a planned combined exercise with the British Royal Marines and the Italian navy on Sardinia. As the situation unfolded in Lebanon and the Marines landed, General Wade eventually took command of all units which had landed.

President Eisenhower's order to deploy the Marines was passed through the chain of command, directing BLT 2/2 to land at 1500 on 15 July on Red Beach. Four miles south of Beirut, just west of Beirut International Airport, and just north of Khaldah, Red Beach would be the scene of another Marine landing nearly 24 years later.

Lieutenant Colonel Hadd's BLT 2/2 landed on time. Before its four rifle companies reached their objective, Beirut International Airport, they were forced to pick their way gingerly through beach obstacles presented by bikini-clad sunbathers and vast numbers of soft drink and ice cream vendors. At the airport, the Marines set up a defense perimeter for the night.

The next day, BLT 3/6 began landing at Red Beach at 0730. At the same time, Hadd's battalion prepared to move into Beirut. After some delays, BLT 2/2 finally left the airport at 1230, and by 1900 had reached the city, where it took control of the dock area and posted security guards around the American Embassy as well as critical bridge sites.

On 18 July, BLT 1/8 landed over Yellow Beach, near Juniyah, four miles north of Beirut. Concurrently, elements of the 2d Battalion, 8th Marines, which had been airlifted from Camp Lejeune by way of Cherry Point, began arriving at Beirut International Airport. Also arriving was the U.S. Army 24th Airborne Brigade which had flown in from Germany and was commanded by Major General Paul D. Adams, who eventually was named Commander in Chief, American Land Forces, comprised of all American troops in Lebanon.

The turmoil in Lebanon settled down after national elections on 31 July. General Fuad Chehab, commander of the Lebanese army, was elected president and on 23 September took office as the head of a coalition government including dissident parties which had been opposed to the previous administration. In mid-August, the first of the Marine BLTs left Beirut, and by 18 October, with the exception of the Security Guard Marines at the American Embassy, all Marines had left Lebanon. While the 2d Marine Division units were in country, there had been only minor confrontations with the Lebanese army and the rebels. A few shots had been fired by both sides, but there were no casualties.

In July 1976, when protracted factional fighting in Beirut threatened the lives and safety of American citizens, Marines were called upon once more, this time to assist in a non-combatant evacuation operation (NEO). The 12-man detachment of the Marine Security Guard at the American Embassy in Beirut, and the naval attache, Marine Colonel Forrest J. Hunt, had radio communication with the evacuating unit, Task Force 61. They controlled the orderly evacuation of 160 American civilians and 148 foreign nationals on 27 July.

Despite efforts of the international community to alleviate the bloodletting in Lebanon, the fighting continued, fluctuating with the fortunes and the strength of each of the factions. The Marines entered

Department of Defense (USMC) Photo A17458

Marines of Battalion Landing Team 2/2 form a LVT and tank column on the beach road for the move into Beirut on 16 July 1958. They were the first of the 2d Provisional Marine Force to enter the city on this date. The Mediterranean is in the background to the left.

Lebanon once again in June 1982, destined to play a larger role than they had ever anticipated.

On 25 May 1982, the 32d MAU, commanded by Colonel James M. Mead, a veteran Marine aviator known as "Large James" because of his height, embarked in the ships of Commodore (Captain, USN) Richard F. White's Amphibious Squadron (Phibron) 4 at Morehead City, North Carolina, for deployment to the Mediterranean as the landing force of the Sixth Fleet.[2] The MAU was comprised of Battalion Landing Team (BLT) 2/8 (Lieutenant Colonel Robert B. Johnston), Marine Medium Helicopter Squadron (HMM) 261 (Lieutenant Colonel Graydon F. Geske), and MAU Service Support Group (MSSG) 32 (Major William H. Barnetson). In all, it had a total of 1,746 Marines and 78 Navy personnel. The BLT included several elements from its parent 2d Marine Division: an artillery battery, a tank platoon, an assault amphibious vehicle (AAV) (amtrac) platoon, a reconnaissance platoon, an antitank (TOW) section, and a communications section.[3]

Like the aviation combat element of other MAUs deployed to the Mediterranean, HMM-261 was a composite squadron. As such, it was comprised of 12 Boeing Vertol CH-46E "Sea Knights," 4 Sikorsky CH-53D "Sea Stallions," 4 Bell AH-IT "Sea Cobras," and 2 Bell UH-1N "Iroquois," more often known as the "Huey."

During the Atlantic crossing, the MAU headquarters continued planning and preparing for a joint amphibious exercise in Portugal, 21-26 June. The staff also began preparing the MAU input to a Task Force 61/Task Force 62 operation order for contingency operations in Lebanon.[4] As the task force neared the Atlantic coast of Spain, however, events in the eastern Mediterranean were combining to disrupt the original deployment schedule. At 0930 on 6 June, the 32d MAU reached Rota, where, at a later date, it would relieve the 34th MAU as the landing force of the Sixth Fleet.*

As Colonel Mead later wrote:

> Within a few hours of tying up at Rota, the message traffic was swelling with stories of the Israeli Defense Force (IDF) attacking into southern Lebanon. At first, the Israeli objectives seemed limited to a 40-kilometer artillery buffer zone in southern Lebanon to protect northern Israel from the shelling of the Palestinian Liberation Organization (PLO). A few days later, the Israelis would attack Syrian surface-to-air missile (SAM) sites in the Bekaa Valley of central Lebanon and eventually encircle the capital city of Beirut in an attempt to destroy the PLO and thereby neuter their political and military influence in the region.[5]

---

*To identify them more clearly as units of II Marine Amphibious Force, the 32d and 34th MAUs were later redesignated 22d and 24th MAUs, respectively. For the 34th MAU, the redesignation took place on 7 July 1982 at Camp Lejeune when it reverted from operational control of the Commander, Sixth Fleet to that of the Commander-in-Chief, Atlantic Fleet/Commander, Second Fleet (CinCLantFlt/2d Flt) and then to Fleet Marine Force, Atlantic, (FMFLant) as noted in 24th MAU SitRep No. 1, dtd 12Jul82. Because it was an element of the Sixth Fleet in July, the 32d MAU did not become the 22d MAU until 1 December, when it was in Camp Lejeune and once again under FMFLant control. Simply put, FMFLant had administrative control of the MAUs while the numbered fleets had operational control.

Anticipating orders to evacuate American citizens from Lebanon, the MAU quickly rearranged the cargo holds of the Phibron 4 ships in order to support the evacuation operations. Also on 6 June, the MAU arrived at Rota where an informal turnover meeting was held with representatives of Phibron 8 and the 34th MAU.[6]

The elements of Task Forces 61 and 62 were placed on immediate alert and steamed out of Rota on 7 June at 0600, proceeding at maximum speed to their designated operating area in the Eastern Mediterranean approximately 100 miles off the coast of Lebanon.[7]

While underway during the next two weeks, the MAU and the Phibron conducted extensive planning and training in preparation for the evacuation of American citizens and foreign nationals from Lebanon. Reporting to Lieutenant General John H. Miller, commanding the Norfolk-based Fleet Marine Force, Atlantic (FMFLant), Colonel Mead stated that he had established effective liaison with the Fleet Marine Officer of the Sixth Fleet, Colonel Jim R. Joy, who was to play a major role in the Marines' Beirut experience. He further stated that, "At this point, we have a grasp of all problem areas and 32d MAU is ready to accomplish its mission."[8] On 17 June, Phibron 4 and 32d MAU officially relieved Phibron 8 and 34th MAU as Task Forces 61 and 62 respectively of the Sixth Fleet.

Prior to their departure from the United States for deployment in the Mediterranean, the MAU and the Phibron had spent a good portion of their time training together practicing the evacuation of civilians from trouble spots. "The procedures were updated by increased training with the TACSIT (*Tactical Situations*) booklet, which provided a series of wargame type scenarios. . . ."[9]

In addition, the 32d MAU staff had reviewed the conduct of Operations Eagle Pull and Frequent Wind, the evacuation of civilians and military from Phom Penh and Saigon in 1975 and had re-read the articles concerning these operations which appeared in the *Marine Corps Gazette*.[10]

On 15 June, the 32d MAU and Phibron 4 were placed on a three-hour alert to prepare for evacuation operations. Nine days later, they received the order to execute. The runways at Beirut International Airport had been heavily shelled and were considered not usable, and the road from Beirut to Damascus ran through the scene of heavy fighting. Accordingly, it was determined that the civilians would be evacuated from the port of Juniyah, approximately five miles northeast of Beirut. Initially, the MAU received reports

A Navy landing craft, utility (LCU) carries civilian evacuees to Amphibious Squadron 4 shipping in the waters off Beirut during the emergency in Lebanon in June 1982.

Using makeshift tabletops, evacuees dine with gusto on Navy rations en route to Cyprus.

that there would be more than 5,000 evacuees, but this figure proved to be quite inflated.[11]

The first landing craft, an LCU, was at the dock in Juniyah at 0800 on 24 June. There was not an evacuee in sight; they had not assembled in Beirut at the designated time. When they did assemble, to aggravate matters, they brought too much luggage. There were not enough buses to carry both the evacuees and all their belongings, but by the end of the day, 580 evacuees had been boated to the ships *Nashville* (LPD 13) and *Hermitage* (LSD 34). Heavily involved in the operation was Major William H. Barnetson's MAU Service Support Group (MSSG) 32, which established an Evacuation Control Center where each evacuee's name and passport number along with other appropriate information were entered into a computer and sent to designated State Department agencies. The MSSG also provided health and comfort items (including toys for the children) from the Landing Force Operational Readiness Material (LFORM), a contingency block that each MAU takes to sea.[12]

Also playing a major role in receiving the evacuees on board the *Nashville* was Commander George A. Gunst, the MAU's Catholic chaplain. Father Gunst was asked to wear his civilian clothes with clerical collar, so that his presence as a minister might have a calming effect on the evacuees, who had been uprooted from their homes in Beirut. Marines gave up their sleeping spaces for the overnight trip to Larnaca, Cyprus. Enroute, they delighted the children with magic shows and other forms of entertainment.[13]

After debarking the civilians at Larnaca, the *Hermitage* and *Nashville* rejoined the other ships of the Phibron. "Morale was sky high as all felt personal satisfaction for helping remove men, women, and children from the real danger of Lebanon where combat actions were intense in many sectors."[14]

On 23 June, Task Forces 61 and 62 were put on a two-hour alert to provide helicopter transportation in support of Department of State peace negotiators. HMM-261 soon became known as the "Cammie (camouflage) Cab Service," shuttling White House Special Envoy Philip C. Habib and Ambassador Morris Draper and their parties, between the flagship *Guam* (LPH 9) and Larnaca, Beirut, and Tel Aviv. During the next four months, the Marine helicopters would fly 62 missions in support of these diplomatic efforts to mediate a peace settlement.

Before long, it became evident that the situation in Lebanon would not be resolved quickly. It also became apparent that the MAU and the Phibron would have to remain on station in the Eastern Mediterranean, not too far from Beirut. Because embarked Marines and sailors had been unable to go ashore for some time, Vice Admiral William H. Rowden, Sixth Fleet commander, sought to relax the alert status to permit some of the ships to visit major ports in the Mediterranean. The *Hermitage, Manitowoc* (LST 1180), and *Saginaw* (LST 1188) left the Phibron for Taranto, Italy, on 1 July for rest and recreation, rejoining the rest of the Phibron six days later. The *Nashville* and *Saginaw* were detached on 19 July for a port visit to Naples. The flagship could not leave the area, however. As a consolation, to celebrate 45 straight days of shipboard time, Marines on the *Guam* "spliced the mainbrace" with two cans of beer per man on 21 June in a special relaxation of the Navy's "no-alcohol-on-board-ship" policy. By 26 July, the political upheaval in the Middle East had relaxed sufficiently to permit the alert status to be increased to 72 hours, and the remaining three ships set course for Naples on 26 July. Before the ships left, two UH-1N helicopters from HMM-261, their crews, and maintenance personnel, were flown to the aircraft carrier *Independence* (CV 62), where they would continue to transport Habib and Draper.

Earlier, when the two diplomats had been flown to and from the *Guam*, they kept Commodore White and Colonel Mead, and their staffs, abreast of events and decisions then underway. Colonel Mead recalled:

> In these discussions, Ambassador Habib and Mr. Draper were most candid with us concerning the progress of peace negotiations. Integral to all their plans and options was the use of our team afloat and ashore.
>
> The discussions . . . focused on utilizing 32d MAU in the roles of disengagement; disarming; destruction of weapons; and assembling, processing, and transport/escort of PLO (Palestinian Liberation Organization). The obvious concerns of inserting some portion of the 32d MAU between 30,000 Israelis and 15,000 PLO and Syrian fighters were well recognized.[15]

While the MAU and the Phibron remained ready in Naples to conduct contingency operations, whatever and wherever they might be, Ambassador Habib continued his shuttle diplomacy between Tel Aviv and Beirut, ". . . seeking a political solution that ultimately required some multi-national force involvement."[16]

Ambassador Habib requested that General Bernard W. Rogers, the Commander in Chief, U.S. Forces, Europe (USCinCEur), provide him with a military liaison team, preferably made up of French-speaking officers, to arrive in Beirut no later than 7 August to assist and advise him with respect to military matters. Lieutenant Colonel Robert B. Johnston, commander

Photo courtesy of Col Robert B. Johnston, USMC

Col James T. Sehulster (left), EUCOM liaison officer to Ambassador Habib, and LtCol Robert B. Johnston, Battalion Landing Team 2/8, 32d Marine Amphibious Unit commander, in front of the U.S. ambassador's residence in Yarze, east Beirut, in August 1982.

of BLT 2/8, was designated the 32d MAU representative, because he was fluent in French, ". . . and while I was a little rusty, I certainly could understand everything and could speak reasonably well. That, plus the fact that I was the BLT commander, a ground officer, really, almost made me the perfect candidate to go in what was called the Habib Mission."[17] Coincidentally, the two representatives General Rogers ordered to Beirut from his headquarters were both Marine officers—Colonel James T. Sehulster, who was to be the liaison team chief, and Lieutenant Colonel Edmond D. Gaucher, Jr., like Lieutenant Colonel Johnston, fluent in French. Together with Lieutenant Colonel Johnston, the team was to undertake the following:

> (1) Provide liaison between Special Envoy Habib, USEUCOM Hq, and the operating forces in planning for executing the deployment of U.S. forces of no larger than battalion size to assist the LAF (Lebanese Armed Forces), and possibly forces from other nations in the Beirut area, after the evacuation of PLO combatants was well underway.
>
> (2) Advise Special Envoy Habib on the feasibility of military operations and employment of U.S. forces. [In this regard the team provided full, straightforward advice to Mr. Habib, but had no inherent decision making authority. Every proposed military tasking was released to Hq EUCOM and passed, as required, to higher authority for decision.]
>
> (3) Establish direct, secure voice communications with Hq EUCOM and the operating forces.

(4) Provide daily situation reports and updates to Hq EUCOM.

(5) Perform such tasks as directed by Hq EUCOM.[18]

Lieutenant Colonel Johnston was briefed in Naples on 6 August and left the same day with the rest of the EUCOM team for the *Forrestal*, then steaming off the coast of Beirut. The next day the Marines were flown to Juniyah instead of Beirut, because the latter was still under Israeli fire. They were met by the resident defense attache in Beirut, Marine Colonel Winchell M. Craig, Jr., who suggested that the liaison team wear civilian clothes instead of uniforms because of a potential sensitivity to a foreign military presence that early in the planning.[19] The team was also required to return each night to the *Forrestal*. As the team returned by helicopter on the evening of the 7th, it was buzzed by two Israeli F-16s, in response to which the United States lodged an official protest. It was to be the first of a number of confrontations the Marines were to have with the Israelis. As Lieutenant Colonel Johnston recalled, "I thought quite frankly the pilots were hotdogging, but as you know, a jet just by [its] sound and speed can blow a helicopter out of the air without shooting at it. So it was of some consequence, I suppose."[20]

When the team landed the next day, it was met by an Israeli colonel who refused to permit the Marines to continue on to Beirut until they identified themselves and stated what the purpose of their trip was. Colonel Craig protested. "And after about 20 minutes, [we] gave them our name, rank, and serial number and indicated simply that we were there to provide support to Ambassador Habib."[21] About 45 minutes later, the team was allowed to continue on its way. During the ensuing days, Ambassador Habib quickly established the organization of the political and military committee, which was to exist for the following two weeks and prior to the entry into Lebanon of a multi-national force. On the military side, there were to be U.S., French, and Italian liaison teams. They did not work independently, but met every day—often more than once a day—in a group session chaired by Ambassador Habib. They discussed the political negotiations concerning the plans for evacuating the

Members of the Military and Political Committees pose before their meeting with Special Envoy Habib in August 1982: 1) American Ambassador Morris Draper; 2) French BGen Jacques Granger; 3) Italian Ambassador Franco Olitteri; 4) U.S. Special Envoy Philip C. Habib; 5) Col James T. Sehulster, USMC; 6) French Ambassador Paul Henri; 7) LtCol Edmund D. Gaucher, Jr., USMC; and 8) LtCol Charles R. Smith, Jr., USMC.

Photo courtesy Col James T. Sehulster, USMC

PLO by sea and/or land, and considered the kind of options that were available.[22]

At the end of each day's meetings, Lieutenant Colonel Johnston briefed Commodore White and Colonel Mead by radio message to keep them fully abreast of what was transpiring in the meetings and how the Phibron and the MAU would potentially be employed in the anticipated withdrawal of the PLO. Lieutenant Colonel Johnston's report also concerned the diplomatic side of the discussions and alerted his superiors to the mission statement that was being prepared by Ambassador Habib, French Ambassador Paul Henri, and Italian Ambassador Franco Olitteri.

The U.S. mission, as it finally evolved after many days of discussion in the military committee, as well as with the government of Lebanon, was to:

> Support Ambassador Habib and the MNF committee in their efforts to have PLO members evacuated from the Beirut area; occupy and secure the port of Beirut in conjunction with the Lebanese Armed Forces; maintain close and continuous contact with other MNF members; and be prepared to withdraw on order.[23]

In Naples, the MAU and the Phibron remained on a 72-hour alert as the discussions in Beirut continued. Although sufficient time was allowed for liberty for all hands, training and preparations for the Lebanon deployment continued. The MAU headquarters also conducted naval gunfire support and close air support communications exercises with respective elements of the Sixth Fleet tasked to provide those missions. At the same time, the BLT trained in fire support coordination and emergency evacuation procedures. HMM-261 conducted gunnery shoots with its Cobras, and technical training for its avionics and corrosion control personnel. And Marines would not be Marines if they didn't participate in physical training daily. As Colonel Mead commented, "Marines and sailors continue to work together in a team spirit that is enhanced by imminent mission."[24]

On 16 August, the Sixth Fleet commander, Vice Admiral Rowden ordered the Phibron to a designated location off the coast of Lebanon, prepared to land embarked Marines on order, perhaps as early as 20 August, as part of the Multi-National Force (MNF). On being given 20 August as the possible L-Day, Commodore White, ". . . requested both a mobile medical augmentation team (an eight-man team which would provide . . . care for surgical emergencies) and a nine-man Environmental Preventative Medicine Unit to assist with field sanitation."[25]

At the same time, Colonel Mead asked FMFLant to augment the MAU with interrogators/translators, ANGLICO (Air and Naval Gunfire Liaison Company) teams, and additional intelligence personnel. All of these elements reported to the MAU within 36 hours after having been requested.[26]

The military committee concluded that the French would go into Beirut first and that the Americans would land only when Ambassador Habib was satisfied that the evacuation was proceeding well and smoothly. The Italians were scheduled to land the day after the Americans. An arbitrary ceiling had been established by Ambassador Habib for the size of the force to be employed—800 French, 800 Americans, and 400 Italians.[27]

Mead and Johnston accommodated to the 800-man ceiling by bringing in, primarily, the rifle companies and the battalion headquarters, and some elements of the MAU headquarters.[28]

The plan called for elements of the Multi-National Force and the Lebanese Armed Forces to be located together at points between the Syrians and the PLO forces in west Beirut, and the Israeli and Lebanese Christian Phalangists deployed in east Beirut. All governments and parties to the plan had agreed to support it, which led ultimately to its successful culmination with but few hitches.

The scheduled day of the landing was slipped because of the inability of the diplomats to achieve a complete ceasefire in Beirut. On 21 August, Lieutenant Colonel Charles R. Smith, Jr., 32d MAU executive officer, was flown from the *Guam* to Juniyah to relieve Lieutenant Colonel Johnston, who returned to the *Guam* the next day, where he resumed command of his BLT. Before leaving Beirut, Lieutenant Colonels Johnston and Smith reconnoitered the port area where the Marines were to be deployed. They selected possible helicopter landing zones, and concluded, based on their observations, that a surface landing of the MAU could be conducted as planned. Upon reaching the *Guam* on the 22d, Lieutenant Colonel Johnston briefed Admiral Rowden, Commordore White, and Colonel Mead, and their staffs.[29]

With H-Hour and L-Day set at 0500 on 25 August, the MAU began final intensive training in such matters as field sanitation, crowd control, and relations with the media, all of which were to be of great concern in the coming days. Field sanitation was especially important in view of the large number of cases of dysentery suffered in the 1958 landing. Because the MAU was landing in Lebanon on a peacekeeping mission in a permissive environment—i.e., one which was not hostile—Colonel Mead decided he would not need his tanks, his artillery, or his attack helicopters ashore at this time. The squadron's transport and utility helicopters would be employed for logistical support.

On the 24th, an advance party from the MAU headquarters flew into the port to reconnoiter the area and to meet with the French, whom the Americans were to relieve. Later that day, Lieutenant Colonel Johnston and his company commanders also conducted a reconnaissance of the port area. That same day, two messages were sent to Colonel Mead on the *Guam*, and relayed to all hands over the loudspeakers of all ships in the Phibron. The first was from Commandant of the Marine Corps General Robert H. Barrow, which read:

> Personal for Colonel Mead from General Barrow:
>
> Please convey the following message to all 32d MAU Marines. You will soon be engaged in carrying out an extremely important mission in Beirut. Clearly, it is also a most difficult and delicate one. Your soldierly virtues, especially discipline, will in all likelihood be severely tested.
>
> At this critical hour you will serve as the primary instrument of our national will to further the course of peace in that troubled region.
>
> As Marines you will meet the challenge and acquit yourselves, not only honorably, but with distinction. The eyes of your countrymen will be on you as surely as their hearts are with you. Beyond that, speaking for myself and your fellow Marines, be assured we have every confidence that as professionals you will superbly represent our Corps and country.

The second message was from President Reagan:

> You are about to embark on a mission of great importance to our nation and the free world. The conditions under which you carry out your vital assignment are, I know, demanding and potentially dangerous. You are tasked to be once again what Marines have been for more than 200 years—peace-makers.
>
> Your role in the Multinational Force—along with that of your French and Italian counterparts—is crucial to achieving the peace that is so desperately needed in this long-tortured city.
>
> I expect that you will perform with the traditional espirt and discipline for which the Marine Corps is renowned. Godspeed. Ronald Reagan

At 0500 on the 25th, the first LCU landed at the port of Beirut and Captain Kenneth T. McCabe's Company E (Reinforced) marched ashore with the MAU colors flying, to be met by a large press contingent, as well as by Ambassador Habib; U.S. Ambassador to Lebanon Robert S. Dillon; the French and Italian ambassadors; the Lebanese Armed Forces' commander, Lieutenant General Victor Khoury; and a host of other dignitaries. Twenty-two minutes later, Captain Richard C. Zilmer's Company F (Reinforced) landed.

At 0600, French Brigadier General Jacques Granger officially turned the port area over to the Marines. On the way in, Colonel Mead noticed the French tricolor

32d Marine Amphibious Unit Marines remain alert while manning a checkpoint during their supervision of the evacuation of the PLO from the port of Beirut in late August 1982.

U.S. Navy Photo by PH3 R. P. Fitzgerald

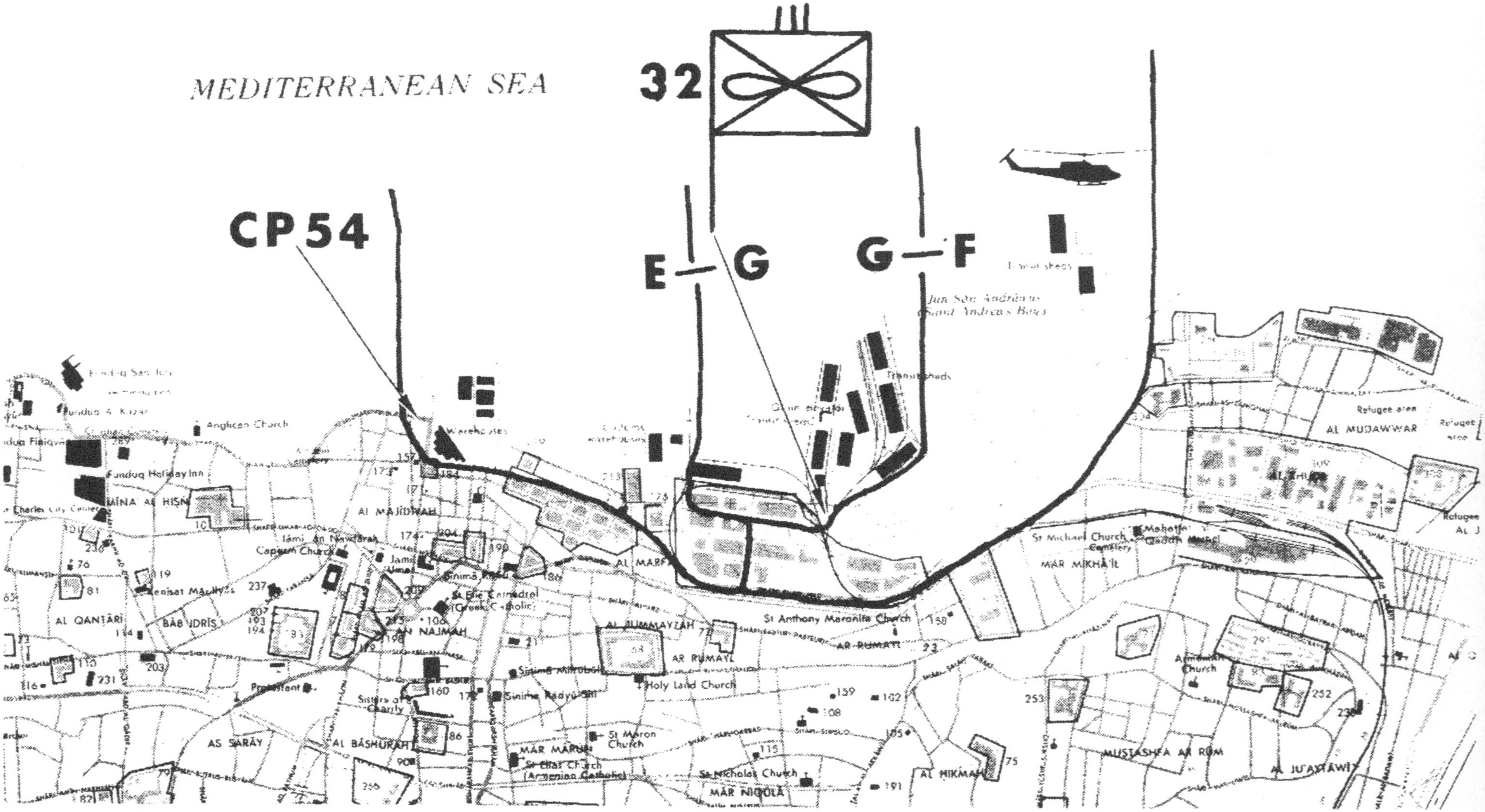

Tactical dispositions of 32d MAU after initial landing at Beirut Port 25 August 1982.

Figure 2

Map courtesy of *Marine Corps Gazette*

Copyright 1983. Richard Allison. *The Army Times Publishing Company.* Reprinted with permission.

flying over the port, and ". . . as soon as we relieved the French . . . the first thing we did was strike their flag and run up the Lebanese flag,"[30] an act that was immediately noticed by the Lebanese and commented upon favorably in the Arab-language press. By 0730, Lieutenant Colonel Johnston had relieved his French Foreign Legion counterpart and accepted command of the port area and all checkpoints therein.[31]

Surrounding the area the MAU occupied was a high fence approximately two kilometers long. Captain McCabe's company was ordered to man strongpoints in the western half of the Marine area, while Company F was similarly deployed in the eastern portion. Captain Alfred J. Karle, Jr., and his Company G were assigned to internal security and to provide a reserve force (See Figure 2). Meanwhile, the headquarters elements of the MAU and the MAU Service Support Group were set up ashore, while the squadron remained on board the *Guam*.

After Company E landed, Captain McCabe inspected his lines and checkpoints, and noted that everybody was settled in and doing a fine job.[32] McCabe tasked his company gunnery sergeant with finding a building for use as company headquarters, which, not too coincidentally, turned out to be the same building the French had occupied. When the French command was relieved, it left one platoon behind at Checkpoint 54, the entrance to the port through which the PLO troops were evacuated.

The Greek ship *Sol Georgious* docked at 0915 that first morning and immediately began taking on board those who were to be evacuated by sea. The Marines checked a total of 1,066 PLO personnel that day. Before the evacuation ended, they were to pass a total of 6,436 Palestinians through their checkpoint.

The following report which Commodore White sent Admiral Rowden vividly describes the departure of the Palestinians from West Beirut:

> My helicopter landed at Landing Zone [LZ] Barrow which is ideally located for ship-to-shore logistics. The LZ is controlled by a Marine team in a most efficient manner. I was met by Col Mead, who took me to the MAU Command Post [CP]. The CP is well set up in a small port office building. Communications are terminated in the building which makes for excellent command and control both among units ashore and with the flagship. The gist of the discussions with Col Mead was that all was well and operating effectively.
>
> At approximately 1000, LtCol Johnston, BLT 2/8 com-

Photo courtesy Maj Fred T. Lash, USMC (Ret)

Marines remain alert as members of the PLO, firing their weapons wildly into the air, leave Beirut headed for the port in trucks provided by the Lebanese Armed Forces.

32d MAU Marines and members of the French 2d Foreign Legion Airborne Battalion form a joint security guard during the evacuation of the PLO from Beirut, Lebanon.

Photo courtesy Maj Fred T. Lash, USMC (Ret)

mander, took me in his command jeep to Checkpoint 54, the entrance to the port used by PLO evacuees. En route we passed at least two ships which had been sunk at the piers (time of sinking unknown). Checkpoint 54 looks out to an area which looks like pictures I have seen of Berlin at the end of World War II. Looking up the street which leads into the checkpoint, I was greeted with the view of a PLO [Syrian?] tank with its gun trained at us at a distance of about 700 yards. (Needless to say, I wore a helmet and body armor during this portion of the tour.) Very soon after my arrival, cars started arriving outside the checkpoint with PLO supporters. During this period the occasional small arms and automatic weapons fire began to increase in frequency and intensity. Shortly thereafter the first series of trucks carrying the PLO fighters came down the street to the accompaniment of gunfire, RPGs [rocket propelled grenades], etc. The fire (all directed skyward) came in some cases from the crowd, the trucks themselves, and for the most part, positions in the rubble I could not see. The trucks belong to the LAF and are driven by LAF drivers.

The trucks containing the PLO are allowed into the sandbagged checkpoint one at a time. They contained males, some females, and children as young as a year or two. There were boys as young as 12 or so dressed in fatigues and armed with automatic weapons. Once inside the checkpoint, the truck is given a very cursory inspection by a Lebanese official (in civilian clothes) along with an LAF officer. The U.S. Marine company commander is present at the truck. The street is blocked by a squad of Marines who also are positioned in and on buildings overlooking the checkpoint. I would estimate 15 LAF soldiers man the checkpoint. When the Lebanese official indicates to the Marine company commander that the truck inspection had been completed successfully, the Marine squad moves aside, and the truck proceeds through and rounds the corner out of sight to await another four truck loads. At this stage, the PLO have their automatic weapons and ammunition in their possession. LtCol Johnston led the first convoy of five trucks to the processing area on the pier. A lieutenant (with me along) led the second convoy. At the pier, the PLO are offloaded and processed through any one of about eight stations manned by the LAF. Processing appears to consist of taking names only. The evacuees then walk to the ship's sterngate (*Sol Georgious* in this case). They enter the vehicle side, leave their weapons and ammo (I could not see into the ships to observe this process). Upon completion, they proceeded up the ramp to the main deck which had awning-covered seats/benches for the trip. Both Marines and LAF are at the embarkation point. One altercation took place while I observed. A PLO fighter did not want to give up his weapon. LAF officers handled the situation forcefully (though not physically other than blocking his way) and well. I did not remain for the complete onload. The ship was underway about an hour after I left.

I can sum up my impressions by saying that the Marines have landed and the situation is well in hand.[33]

The MAU went ashore armed with unloaded individual and crew-served weapons, for a deliberate decision was made to demonstrate that the Americans were on a peace-keeping mission. Additionally, they had to show that they trusted the Lebanese Armed Forces to maintain security. The absence of a single military commander in charge of the Multi-National Force created problems in coordination for the respective MNF commanders. This situation prevailed despite the fact that the French and Americans—and the Italians after they landed on 26 September—exchanged liaison teams.[34]

As an example of the confusion that existed, on several occasions French and Italian military convoys passed through the western port area at the same time PLO evacuation convoys were being escorted to the docks.[35] Another problem which arose early was that of identifying friendly forces, as most of them, as well as the Lebanese Armed Forces, Israelis, and some of the PLO, like the Marines, wore camouflage utilities. The BLT later recommended that each force deployed to a similar type of operation be issued a contingency block of U.S. flags which could be flown on vehicles and "brassards with attached flag" to be worn by U.S. forces ashore.[36]

EUCOM peacetime rules of engagement dictated that the Marines were to carry unloaded weapons, although it does not take long to insert a magazine into a weapon and chamber a round. The rules also dictated that the on-the-scene commander had the right to determine what the appropriate response would be if there was a hostile act committed against Marines. Primarily, it would be rifle against rifle, and the like. "The inherent right of self-defense prevailed."[37] As far as Colonel Mead was concerned, the Marines were ". . . comfortable with our rules of engagement, which had been kept simple and therefore readily understandable.[38]

The living conditions in the Beirut port area were rudimentary at best. Most of the units lived in buildings which had been damaged six years earlier in the civil fighting of 1976, and never repaired. Being a port area, it was infested with many rats. The Navy preventative medicine unit worked full time to reduce the infestation, but it was a losing war. Potable water was not available, so it had to be brought in from Phibron shipping, which also provided fresh fruit and sandwiches to the Marines ashore, and thus added variety to their diet of C-rations.

While the Marines had instructions not to deal directly with the Israelis, the MAU operations officer, Lieutenant Colonel Dennis R. Blankenship, did have a hot line to the Israeli liaison officer who was situated in a 15-story structure, the Electric Building, which, though outside the port area, had very good observation of the evacuation proceedings.[39]

The Israelis frequently disrupted the evacuation

operation by blocking the port entrance, refusing entry to commercial shipping coming in to remove the PLO. Israeli gunboats often held the ships outside the port until noon. This created a backup of the PLO, which was being formed and organized in groups of 1,000 for the exodus, and shooting wildly into the air all the while. "The longer you held them, the greater potential for problems existed."[40]

During the entire course of the evacuation, Lieutenant Colonel Johnston was present at the dock, directing it.

> I always made a point of standing right at the ship. I would actually bring the first group of trucks, they had 17 vehicles in all, and they would come in groups of five into the checkpoint. When they were behind the checkpoint within the Marines' positions and we were sure they had no contraband, RPGs, large weapons, my vehicle would guide them to the embarkation point and I stayed there and subsequent convoys came through. So, I was really standing there observing every PLO go aboard the civilian ships.[41]

The highlight of the evacuation was the departure of PLO leader Yasser Arafat from Beirut. There was some question of whether he was going by air or by ship, and given the potential volatility of the situation, exactly when and how he was to depart was very closely held. On 29 August, the day before he was to leave from the port, the Marines were informed of his impending departure. Accordingly, on the 30th at 0500, the BLT made a final security sweep. By 1000, the *Atlantis*, the ship that was to carry him out, had docked, and was ready for Arafat's scheduled appearance at 1100.

According to the plan, he was to be driven to Checkpoint 54 by car, instead of by truck, and from there escorted by the Marines to the evacuation point, about 3/4ths of a mile away. Although his time of departure was supposed to have been a secret, by 1000 a large crowd of media, well-wishers, and hundreds of PLO dependents were on hand to witness the departure. When he neared the checkpoint, Arafat got out of his car to accept the flags of several of his PLO units. His entourage was ". . . led in by the French ambassador, Paul Henri, and a contingent of French troops with armored carriers; a truck full of troops, and they jumped out on their side of the checkpoint as though they were protecting him and making sure there were no snipers."[42] Lieutenant Colonel Johnston stood in front of the checkpoint, preventing the entourage from proceeding further, and had fairly lengthy discussion with Henri, asking him why he thought the French troops were needed when there were 800 Marines present who were perfectly capable of seeing Arafat to the ship that was waiting for him. According to Lieutenant Colonel Johnston, Ambassador Henri replied that Ambassador Draper had agreed to the presence of French troops at the evacuation. Johnston then spoke to Draper over his walkie-talkie, saying that he knew nothing of this agreement and that the exessive number of French forces were creating a problem.[43] It appeared to Lieutenant Colonel Johnston that the

Surrounded by reporters and his security guards, Secretary of Defense Caspar W. Weinberger, joins 32d Marine Amphibious Unit Marines in dining al fresco *on C-rations.*

Photo courtesy Maj Fred T. Lash, USMC (Ret)

French were providing Arafat a guard of honor in the midst of a highly combustible situation.

Present at this confrontation in addition to Johnston was Colonel Mead, Captain McCabe, and a squad of Marines, plus a second squad, all whom physically pushed the crowd back. A concession was made to let some of the French vehicles through, but the Marines were determined that the French ". . . were not going to lead Arafat . . ." into the port area.[44]

Arafat's vehicle entered the port area first, escorted by about 25 PLO bodyguards, all of whom appeared concerned that they were going to relinquish protection of their chief to the Marines. The PLO slapped their rifles and made threatening gestures to the Marines, and actually attempted to push themselves through the Marine checkpoint. The Marines in turn pushed the bodyguards back. Adding to the tension was an accidental discharge by one of the French troops. Fortunately nothing happened.

Observing the evacuation operation at the port was Colonel Craig, who sent the following message to the Commandant:

> I was on hand today at the checkpoint manned by Echo Company, 8th Marines. When Arafat came through enroute to his departure, they handled themselves with a coolness and professionalism that I have seldom seen in my nearly 30 years' experience. They were calm under the pressures of a situation that could have been a disaster. Confrontation occurred with the French as well as a mob of Palestinian admirers of Arafat, most of them armed and excited. The resulting peaceful solution was a credit to Captain McCabe, Lieutenant Colonel Johnston, and MAU commander, Colonel Mead.
>
> It was evident in my hour-long visit that officers and NCOs knew what they were doing and were able to convey their intent to the Marines in their charge. The troops were disciplined and responded smartly. At no time did I see any Marine behaving in any manner but with full concentration on the tactical situation at hand. They were a credit to their country and their Corps.[45]

After this, the evacuation operation continued on *pro forma* with the Marines maintaining port security. By 3 September, Colonel Mead noted a significant change in the atmosphere. In Beirut, the lights were on again at night with restoration of the city's power plant. There was increased traffic in the streets, shops were reopening, and the Lebanese were seen repairing damaged buildings and cleaning up the rubble. As of the 3d, all PLO and Syrians had been evacuated and the Marines began to be visited by such luminaries as Secretary of Defense Caspar W. Weinberger, Senator Charles H. Percy, and Congressmen Edward J. Derwinski and Stephen J. Solarz, as well as the Embassy's new Deputy Chief of Mission, Robert L. Pugh, a former Marine.

There was much work for the MSSG, as it began repairing vehicles and doing preventive maintenance on them, purifying water, and providing other logistical services, such as distributing rations. The squadron was also kept busy flying Ambassador Habib and his associates on "shuttle diplomacy" missions as well as ferrying visiting VIPs from ship to shore and back, and from either ship or shore to Larnaca.

Initially, the MAU staff thought that once its port security assignment was completed, it would man checkpoints on the Green Line, which geographically and historically separated Muslim west Beirut from Christian east Beirut and ran south to a point below the port area. Lieutenant Colonel Johnston and his BLT 2/8 staff visited the checkpoints they thought they would have to man. However, this mission never materialized and the MAU remained in the port enclave until it began reloading on board Phibron shipping on 9 September.

In their after-action reports of this evacuation operation, Colonel Mead and his subordinate commanders made a number of recommendations and noted several lessons learned. Notable was the assertion that, "The operation was distinctively unique from virtually any point of view. Most significant, from a military vantage point, was the fact that military forces from the United States, France, Italy, Lebanon, Syria, and Israel, as well as the PLO forces themselves, cooperated in the relatively small geographic area of Beirut, without the benefit of a Combined or United Command Military Headquarters to direct the operation."[46]

The MAU found that augmentation personnel, who were flown to Beirut upon request of Colonel Mead and Commodore White, often arrived without their personal records and personal field equipment, such as helmets, web gear, weapons, and the like. Additionally, in response to a request for essential elements of information required by the MAU commander to prepare his estimate of the situation ashore, Colonel Mead was given very little tactical intelligence concerning the location of sites of weapons which might have placed indirect fire on the Marines had the permissive environment in Beirut changed. Accordingly, Colonel Mead recommended that, should the MAU face a similar deployment ashore, it should be provided with accurate and timely target intelligence.[47]

Commenting on BLT operations, Lieutenant Colonel Johnston noted that his planning process was greatly hampered by the lack of a current port and

Beautiful Martyr Square, in west Beirut, before 1976.

Martyr Square in 1983. Marines of 24th MAU began patrolling in the ruins here on 4 November 1982.

Photo courtesy Lt Paul E. L. Holdom, Royal Marines

32d MAU Marines board their LST on 10 September 1982 following the PLO evacuation from Beirut.

beach study for Beirut.[48] He also noted that his BLT Marines conducted the entire evacuation operation without inserting magazines or chambering rounds, and there were no accidental discharges. However, he also reported that, "During critical periods, selected marksmen had magazines inserted, rounds chambered, and were ready to engage any threat."[49] Lieutenant Colonel Johnston commented favorably on the enthusiastic support and cooperation of the Phibron and its beachmasters in supporting the Marines ashore with such services as regular mail, laundry, bag lunches, fresh fruit, pay call, and religious services. Finally, the BLT commander remarked that the success of his BLT benefitted from the fact that half of his troops were on a second Mediterranean deployment.[50]

Lieutenant Colonel Geske's squadron played an important role in the operation, providing logistical support to the MAU, to the Phibron, and to the diplomatic mission in flying its various members about the Middle East. In addition, the Cobras were kept on standby in case they were needed for close-in fire support. Later, CH-53s and CH-46s often had to make overwater logistic support flights of up to 150 miles one way, generally to Larnaca. These flights were conducted safely, supported by shipboard radar on board the large carriers located below the horizon at a distance from Beirut, and guided by E-2C airborne control aircraft.[51]

The success of the MAU was further recognized on 10 September, when both the President and the Secretary of Defense telephoned Colonel Mead to compliment him on the performance of the Marines as part of the Multi-National Force.[52] Said Colonel Mead after these calls, "only then did it dawn on me that I had received only three orders during the whole operation. There were to go in, to stay off the Green Line, and to come out!"[53]

After having withstood the pressures of international attention, the MAU Marines and the Phibron sailors exhibited a certain air of self-satisfaction for a job well done as they sailed from Beirut on 10 September. They could not be faulted. Nor, at this time, could they anticipate that they would be revisiting Beirut shortly.

CHAPTER 3

Beirut II—The Mission of Presence 29 September-1 November 1982

As the MARG ships bearing the Marines to the Italian ports of Naples and Taormina neared their destinations—where there would be liberty for all personnel, interrupted only by normal training and necessary ship's maintenance—the MAU and Phibron staffs prepared for their next task—conducting Operation Display Determination 82 in Saros Bay, Turkey, on 25 September. The Marines also spruced up for an impending visit by their Commandant, General Barrow.

A day away from Italy, however, on 14 September, all hands were shocked to learn of the assassination of President-Elect Bashir Gemayel, who, just five days earlier, had reviewed a combined MNF honor guard and had visited with Colonel Mead. The Americans perceived that a return commitment to Beirut was imminent. This perception was sharpened by the news of the massacres on 16 September in the Palestinian refugee camps of Sabra and Shatila in west Beirut.

These two developments impelled the new President, Amin Gemayel, brother of the slain Bashir, to request the return of the Multi-National Force to ensure the safety of the population of west Beirut until the Lebanese Armed Forces were able to undertake this mission on their own. In addition, the MNF presence in Beirut would facilitate the withdrawal of Israeli forces from Beirut to the south and east.[1]

The MARG was ordered to return to the eastern Mediterranean to await further orders, and early in the morning of 22 September, it steamed from Naples (*Manitowoc* from Taormina) for the waters off Beirut. The next day, 32d MAU Executive Officer Lieutenant Colonel Smith was flown from the *Guam* to Larnaca and then on to Beirut to function once more as the MAU representative to the Multi-National Force Liaison Committee, which again included the two Marines from EUCOM, Colonel Sehulster and Lieutenant Colonel Gaucher.[2]

While underway, the MARG was visited on 26-27 September by Vice Admiral Rowden and General Barrow, who in a ceremony on the *Guam*, presented Navy Unit Commendations to Phibron 4 and the 32d MAU for their performance during the PLO evacuation.[3] During the course of the short sea trip to Beirut, the Sixth Fleet was alerted to the possible participation of the 32d MAU in a Lebanon Multi-National Force peacekeeping mission.

Of particular interest in the document establishing the U.S. peacekeeping mission was the statement concerning rules of engagement. The 32d MAU was told that if its assigned area or lines were infiltrated by units other than those of the Lebanese Armed Forces, the intruders were to be warned that they were in an unauthorized area and could proceed no further. If the intruders failed to withdraw, the MAU commander was to be informed of the incident and would decide what further action was to be taken. Only if Marines and any accompanying Lebanese Armed Forces were fired upon, could the Marines return fire to insure their safety and that of the Lebanese. Finally, the Sixth Fleet commander was to be prepared to extract American forces from Beirut, if it became necessary.[4]

The question of the MAU providing other than internal security (i.e., other than for its own positions) was addressed early on and had to be faced as the Marines' mission unfolded. Because of the low threat initially confronting the Marines in Lebanon, the anticipated early capability of the Lebanese Armed Forces to provide security, and the provision in the initial deployment order to be prepared to withdraw when directed, there was no need to change the overall mission, the concept of operations, or the rules of engagement until September 1983.[5] As the Long Commission noted later, there ". . . was no perceived need to change the USMNF role in response to the development of a combat situation, since USCinCEur had been tasked to protect U.S. forces and, on order, to be prepared to extract U.S. forces. Under the circumstances of the combat resulting for USMNF personnel, NCA [National Command Authority, i.e., the President] would be consulted concerning withdrawal."[6]

The commander of each MAU deployed to Beirut made certain all his troops were thoroughly briefed and fully cognizant of the rule of engagement.

Meanwhile, Ambassador Habib was once more in the midst of negotiations and in a most difficult and trying mission—attempting to develop a plan agreeable to the governments of Lebanon, Israel, France, and Italy. From this plan evolved the 32d MAU's mission which was to provide ". . . a presence in Beirut, that would in turn help establish the stability necessary for the Lebanese government to regain control of

Copyright 1983. Dwane Powell. *Raleigh News and Observer.* Reprinted with the permission of the Los Angeles Times Syndicate.

their capital. This mission required the 32d MAU to occupy positions in the vicinity of Beirut International Airport and establish and maintain close continuous liaison with the French, Italian, and Lebanese forces."[7]

During the 18 months in which Marines were deployed to Lebanon, this mission was not much changed. As the Long Commission later concluded: "The 'presence' mission was not interpreted the same by all levels of the chain of command and perceptual differences regarding the mission, including the responsibility of the USMNF [U.S. multi-national force; the Marine Amphibious Unit] for the security of Beirut International Airport, should have been recognized and corrected by the chain of command."[8]

It was basically assumed that the Marines were going into a permissive environment, and for that reason, the mission, rules of engagement, and concept of operations, as well as force structure, were designed to maintain a balance between political and military considerations and requirements. The MAU's location at the airport was considered to be the least threatened position of those held by MNF units and of serving both political and military necessities. The French and Italians would occupy far more dangerous areas, for the former were to be located in the city of Beirut proper, with its narrow streets, teeming rabbit warren-like alleys, and always-heavy traffic, the best possible locale for terrorist activities. Even more ominous was the area assigned to the Italians, south of the city. It was chock-a-block with the ruins of the still heavily occupied refugee camps of Sabra, Shatila, and Burj al Barajinah, where smoldering hate, resentment, and hopelessness lingered.

As the MAU once again neared Beirut, the American mission of "presence" was repeatedly discussed and analyzed by Colonel Mead and his staff. The concept of "presence," as such, was not taught in any of the military schools Marines have attended. Perhaps closest to this concept was the commonly accepted tenet in international law of "interposition," which results when a major power provides military assistance in the form of troops at the request of a legally constituted and established government unable to protect foreign citizens and property. Certainly that was the basis for Marine presence in Central American countries during the 1920s and 1930s, Lebanon in 1958, and Santo Domingo more recently in 1965. However, this was not quite the case in 1982, and Colonel Mead determined that once ashore, he would establish hasty defenses at the airport and collocate his troops as soon as possible with the Lebanese Armed Forces to demonstrate visibly U.S. intentions to support the government of Lebanon.

Beirut II, as this second deployment was dubbed, would be considerably unlike Beirut I. In positions at the airport, the Marines would be in the midst of an area densely populated with Shiite Muslims, who had close religious ties with Iran and venerated the Ayatolla Khomeini. The landing force would be expanded to 1,200 Marines, who, while landing in a friendly environment, would still face the dual threat of individual acts of terrorism and a considerable number of unexploded munitions.[9]

The airport area, scene of heavy fighting, was littered with ". . . literally tens of thousands of pieces of unexploded munitions of 125 types from 19 different countries that had accumulated over the previous 8 years of fighting."[10] Since the earlier personnel augmentation had returned to stateside units, Colonel Mead called once again upon FMFLant to provide him with combat engineer, interrogator/translator, explosive ordnance disposal, public affairs, preventative medicine, ANGLICO, and intelligence detachments. Again, it took less than 36 hours for these to join the MAU.

In conducting a map reconnaissance of the positions he was to occupy at the airport, Colonel Mead soon determined that he needed the high ground approximately five kilometers east of the airport to guarantee the MAU's safety should the situation ashore begin to deteriorate. However militarily sound the rationale for this plan, Colonel Mead soon learned the political realities of life in the Middle East. In the diplomatic discussions leading to the reinsertion of the Multi-National Force, Ambassador Draper and Israeli Minister of Defense Ariel Sharon mutually agreed to permit the Israeli forces to use the Old Sidon Road for resupplying their troops in the Shouf mountains east of the airport. The road just about paralleled the eastern perimeter of the airport and ran southwest to northeast between the airport and below the heights Colonel Mead wanted to occupy. If the Marines took up positions there, it would create a politically unacceptable perception that the United States was protecting Israeli resupply routes. Thus the Americans would be looked upon as anti-Muslim and certainly less than neutral with respect to the Israelis—an image completely opposite from that which the Marines wanted to portray.

In the week between the time Task Forces 61/62 left Naples and the day they arrived in Beirut, a large joint service public affairs team, headed by a Navy captain from EUCOM, arrived to help the MAU and Phibron with press relations. Members of the 12-man team were sent to each of the Phibron ships to brief the Marines and sailors on what to expect from the large media representation expected at Beirut, and how to answer questions.[11]

In addition to briefing the Marines on press relations, the BLT operations officer, Major Raymond Cole, prepared and videotaped an orientation lecture to be shown on all the Phibron ships to all Marines. In this presentation, he discussed the upcoming mission, what it would be like to go ". . . into that kind of environment, some of the do's and don'ts, [while undertaking] permissive operations. . . ."[12]

Of prime consideration at this time is the fact that the rules of engagement received from higher headquarters were written in such constricted legalese that it was necessary to simplify them so that the lowest ranking Marine could readily and thoroughly understand what they meant. Said Lieutenant Colonel Blankenship, "Basically, it was minimal force necessary . . . we did not have magazines in, that was the decision that was made. Fortunately, as it turned out, it was a very right decision." He continued, "Minimal force necessary and we did have a right to self-protection . . . we told our Marines, 'If a guy shoots at you and you feel it is directed fire . . . then you use minimal force necessary to take care of the situation. If he is shooting at you with small arms, you can return the fire with small arms. . . . Just because a man is shooting at you with small arms doesn't mean that you can call in naval gunfire on them or a flight of F-14s with napalm or something like that."[13]

One of the preconditions for landing the Marines in September was that all Israeli forces had to be south of the airport and out of the MAU's assigned area of responsibility. Because this condition was not met on time, the Marine landing was delayed until mid-day, 29 September. The original MAU plan called for a typical Marine landing—two companies via air and one via surface with all supporting arms—into the vicinity of Beirut International Airport over what was then called Black Beach and subsequently renamed Green Beach.[14] Then the Marines would push forward, clearing the airfield to the foothills, which included the high ground at Kfar Shima and Shuwayfat, tying in with the Presidential Palace and the Ministry of Defense in the Baadba area.[15]

Major Jack L. Farmer, Assistant MAU Operations Officer in Beiruts I and II, recalled that the staff was concerned whether its plan would work, for in considering the MAU's strength at the time, Marine lines would be stretched out very thinly if the mission was a tactical one of defense against a hostile force. That the mission was diplomatic rather than tactical meant that reinforcement by an additional BLT was unnecessary. The MAU determined that it could accomplish its mission with the BLT and the MSSG augmentation it already owned, and that it had enough Marines to establish strong-points along a general trace going from Baabda, through Kfar Shima, Shuywayfat, and to the south of the airport at Khaldah. Additionally, the MAU would begin foot and motorized patrols to satisfy the requirements of its presence mission. Based upon political considerations and the diplomatic climate at the time, the MAU did not want to give the impression that it was securing the Old Sidon Road to protect the Israeli Force. As a result, the MAU reduced its perimeter by settling in west of the road and pulling in closer to the airport.[16]

It was decided to restrict the Marines to a location in the vicinity of and to the west of the unused railroad tracks at the southeastern portion of the airport perimeter, further reducing its lines. Actually, the reduction gave the MAU better internal lines of communication but it did raise subsequent questions by the Israelis about where the Marine lines actually were. The military importance of the high ground to the east of the Old Sidon Road was reiterated to Ambassador Habib and his diplomatic assistants by Lieutenant Colonel Smith, who conducted a reconnaissance of the general area together with the

U.S. Navy Photo

32d MAU commander Col James M. Mead watches as his Marines return to Lebanon on 29 September 1982 on board their LVTP-7 amphibious vehicles.

A 32d MAU troop leader posts his Marines on the edge of Beirut International Airport, on the outskirts of Burj al Barajinah refugee camp shortly after their return to Lebanon.

Photo courtesy of Claude Salhani

U.S. Navy Photo

Capt Richard C. Zilmer leads his Company F, Battalion Landing Team 2/8 Marines ashore from the landing ship Saginaw *(LST 1188) at the port of Beirut on 29 September 1982.*

embassy staff.[17] While Habib may have recognized the validity of the Marines' rationale to hold the heights, he still refused to permit the MAU to occupy them.

The territory the MAU was to occupy on the eastern portion of the perimeter also held built-up areas, in which there were located one Christian-oriented village—Kfar Shima—and one pro-Druze/Muslim village—Shuwayfat. Subsequently, after December 1982, these two villages were the locale of a considerable amount of the factional fighting of the period. Major Farmer believed that proximity of the villages to the Marine lines may have caused the MAU problems of internal security. On the other hand, he concluded, ". . . our presence there may well have been a stabilizing factor which would have allowed a more comprehensive solution, at least in the context of those two villages," and would have prevented the problems which finally arose with the Israelis in that area getting caught in ambushes along the Old Sidon Road, resulting in subsequent confrontations between the Marines and the Israelis.[18]

Putting plans into action, at 1158 local time* on the 29th of September, the *Manitowoc* tied up at the dock in the port of Beirut and Colonel Mead landed with his staff and Company F. The Marines were met by Ambassador Dillon, Lebanese officials, and a horde of media.[19] The *Saginaw* docked at 1340 to unload the vehicles which would carry the Marines to the airport, and at 1400, the first of the helicopter-borne troops landed at the airport. By 1700, all three rifle companies and other scheduled personnel and equipment were ashore. All organic firepower was brought ashore with the exception of artillery and tanks, which remained on board their shipping for the entire duration of Beirut II.

The next day, 30 September, the Secretary of Defense released the following message from the President:

> At the request of the Government of Lebanon, I have ordered the Landing Force, Sixth Fleet, to return to Lebanon as a part of the Multi-National Force. I well recognize the requirements and demands that this places upon you—the members of the fleet. I also know the steadfastness and devotion to duty you have displayed throughout the ordeal of this tortured land. The cause of peace—and the interests of our nation—are being well served by all of you who go down to the sea in ships. Be assured that you have the unending gratitude of all who love freedom. God Bless You. Ronald Reagan.[20]

Two other events of note occurred on the 30th. The

*Unless otherwise noted, all times stated in this monograph will be local time.

USMC Photo by SSgt Robert E. Kline

Green Beach is where Marines and their equipment landed in Lebanon first during the 1958 emergency and then 14 years later, when they gave "presence" in Beirut, 1982-1984.

first was a Multi-National Force welcoming ceremony when President Amin Gemayel reviewed and spoke to representative groups of U.S., French, and Italian troops.* The second event was sobering for it marked the first Marine death in Lebanon. From the time of its landing in country on 29 September and throughout its first deployment, the 32d MAU continually conducted an ordnance and disposal operation. While engaged in clearing the airfield, Corporal David L. Reagan was killed, and three other Marine engineers wounded, when a bomblet exploded. The casualties were quickly evacuated to the *Guam*, where they were treated by a special Camp Lejeune medical team attached to Phibron 4.

Also on 30 September, Green Beach was cleared of mines, but not completely of the bomblets. Nevertheless, Colonel Mead ordered his amtracs ashore as well as the Service Support Group and the Navy beachmaster unit. The next day, the rifle companies began moving into their assigned positions to the south and east of the airport, while the Marines in the northernmost positions linked up with the Italians. An analysis of the terrain indicated that the 32d MAU should establish some forward observation and listening posts collocated with those manned by the Lebanese Armed Forces. The MAU also decided to dig positions to the rear of these posts, closer to the airport and the runways, to be occupied when and if Marines were attacked by a hostile force.[21]

The forward posts were located at key intersections of the nearby road network surrounding the airport—primarily those access roads which ran from the Old Sidon Road to the airport. A small string of checkpoints was located due east of the airport, with the

*The initial French input to the Multi-National Force was comprised of a staff from the 11th Airborne Division and units from the 9th Marine Infantry Division, 2d Foreign Legion Airborne Battalion, 3d Marine Airborne Division, 9th Headquarters Support Battalion, and 17th Airborne Engineers Division, all commanded by Brigadier General Jacques Granger. The Italian unit initially deployed in Beirut was the 2d Bersaglieri Battalion "Governolo," which returned on 27 September with 300 paratroopers from the "Folgore" Airborne Brigade (two companies from the 1st Carabinieri Airborne Battalion "Tuscania," and one company from the 9th Airborne Assault Battalion "Col Moschin"). These units were reinforced by additional personnel from the San Marcos Naval Infantry Battalion, commanded by Commander Pierluigi Sambo, Italian Navy. Both Commander Sambo and his battalion were well known and respected by Marines who had conducted combined landing exercises in the Mediterranean with the Italians. In overall command of the Italian MNF unit was Brigadier General Franco Angioni. For a complete order of battle of the foreign MNF units, see Appendix C.

U.S. Navy Photo

Marines sweep the Green Beach area for mines and unexploded ordnance to secure it prior to the landing of the rest of the 32d MAU in September 1982.

first post in the south subsequently designated Checkpoint 76. It was located on the access road that ran from Old Sidon Road to the airport past a Pepsi Cola bottling plant. Three other Marine posts were located with Lebanese Armed Forces positions, which ran all the way up north to the vicinity of the small Shiite village of Hay es Salaam (called "Hooterville" by the Marines) and on to Lebanese University, where the MAU had its forwardmost-deployed company set up in a building on campus. Here the Marines would be involved in civic action projects as well as giving "presence." As Major Farmer recalled:

> It was along this section of our perimeter that we had the most problems with the Israeli Defense Forces, especially when [beginning in December 1982] [Israeli] convoys [were] ambushed on the Old Sidon Road and they would return fire.[22]

Southwest and west of the university were a string of outposts beginning at the airport itself. They were lightly manned due to the fact that they were near the Palestinian refugee village of Burj al Barajinah, a largely built-up area which did not lend itself to the establishment of forward outposts. Besides, that sector of the Marine area of operation was partially patrolled, and, in some cases, manned by the Italians at static checkpoints. The MAU maintained close coordination with the Italian-manned checkpoints, which were actually strong points located on the northern perimeter itself. After the Marine artillery was landed in a later MAU deployment,[23] battery positions were set up in the northern perimeter in an area originally controlled by the Italians. There were also several U.S.-Lebanese checkpoints on the Beirut-airport terminal highway, including the main circle road next to the Mid-East Airlines building.

Further to the west was a randomly manned Marine position at the northern end of the north-south runway, and still another one located on the beach, where the MSSG landing support party and the beachmaster unit were located. These positions were maintained during the entire Marine stay in Lebanon. Between the beach position and the airport was a coastal highway which ran from Beirut down through Khaldah and then into the Old Sidon Road, going down to the city of Sidon in the south. The MAU could not cut off traffic on this highway or the airport highway without first coordinating with the Lebanese Armed Forces, for a disruption of these two routes would cause the Lebanese government domestic problems. However, during high threat situations—such as the terrorist attack on an Italian motorized patrol later in the deployment as well as a grenade attack on one of the Marine foot patrols—the MAU did close the roads.

One highly visible post, really an interior guard post, was established at the head of the road leading off from the airport highway and past the MAU headquarters and the MSSG area down to the airport facilities. "We later named it 'Fort Apache' because of the design. Subsequently, sandbagged tar barrels were erected in April 1983, at the time of the embassy bombing, to counter terrorist threats."[24]

The Marines dug in in the conventional manner at the posts established around the perimeter. "I think we filled some 200,000 sandbags in 30 days, and built some pretty fancy defensive positions, particularly on the southern part of the airstrip. . . ."[25]

The MAU headquarters itself was located in the administrative area of the airport. Beirut had an active international airport which, in the two-week period prior to the 23 October 1983 bombing of the BLT building, serviced an average of 35 flights and 2,400 passengers daily.[26] Approximately 1,000 civilians were employed at the airport at this time and some 3,000 civilian and Lebanese military vehicles entered and left the BIA area every day.

The headquarters of the 32d MAU, and of those MAUs that succeeded it, was located in a two-story reinforced concrete building, which had formerly

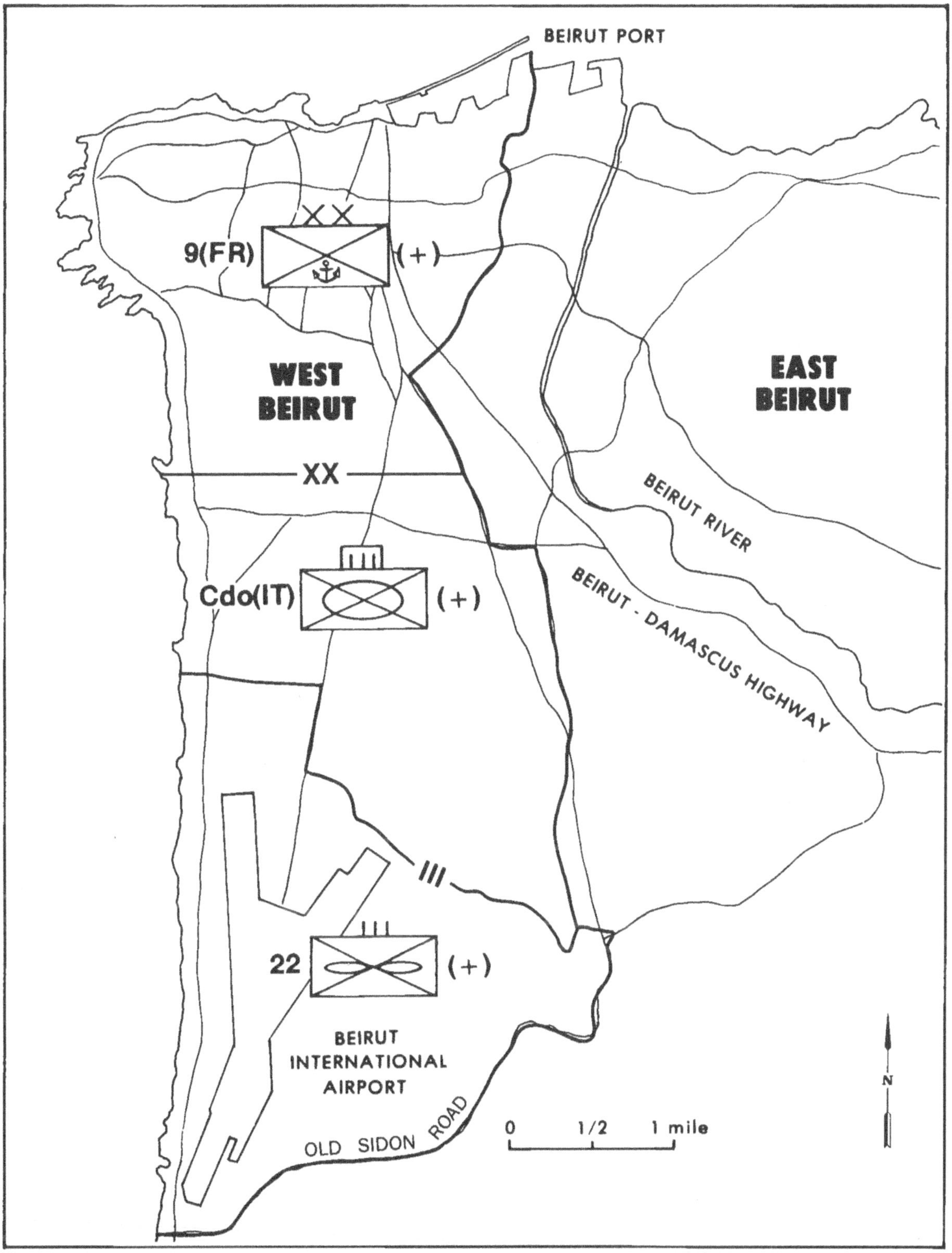

Sketch map courtesy *Marine Corps Gazette*

The zones of the Multi-National Force in September 1982.

USMC Photo by Sgt Kevin D. Dewey

Aerial view of the Beirut International Airport administrative building used to house the Marine Amphibious Unit headquarters in 1982-1983. Note sandbagged vehicle bays.

housed the airport's fire fighting school facilities. The ground floor of the building held vehicle bays, some offices, and an utility room. The MAU commander's office and sleeping quarters were on the second floor. The ground-floor vehicle bays had metal doors and served initially as sleeping quarters for MAU staff officers and staff NCOs. A ground floor room at the end of the building was used as a club for the officers, staff noncommissioned officers, visiting journalists, and VIPs. During Colonel Mead's two tours in Lebanon, this facility was dubbed "Large James' Tavern."[27]

The windows of the second floor offices had been blown out during the earlier fighting and they, as well as all exposed openings, were protected by sandbag walls. The roof, which could be reached by an exterior ladder, served the Marines as an antenna farm.

Immediately across the road from the MAU headquarters was the headquarters of the MSSG. It occupied a single-story, steel reinforced concrete building, whose exposed openings were also protected by sandbags.

The BLT occupied a bombed-out, fire-damaged, four-story reinforced concrete building, southwest of the MAU headquarters. Before the Israeli invasion, the exteriors of the second through the fourth floors held large plate glass windows. By the time the Marines arrived, all windows had been damaged or blown out, and when the BLT Marines moved in, they filled the windowless gaps with an assortment of plywood, plastic sheeting, screen wire, and sandbags. The ground floor was a large open area, which the Marines enclosed with an extensive amount of barbed wire and sand bags. In the center of the building was an atrium, which in turn, was covered with louvered panels that allowed cooling and illumination as well as protection from the elements. Concrete stairwells were at the east and west ends of the inner court. This building had been successively occupied by the Government of Lebanon Aviation Administration Bureau, the PLO, and the Syrians, who used it as a hospital. When BLT 2/8 landed in September 1982, it set up its command post in this structure.

Along the airport road fence immediately to the west of the BLT building, several guard posts were established and sandbagged in. Along the fence also were two amphibious assault vehicles (AAVs or LVT-7s) which were used together as a mechanized command post by Lieutenant Colonel Johnston and successive BLT commanders. South of the BLT building was a blacktopped parking lot, where there was overflow parking for travelers and airport employees. The Marines took up the northermost half of this area as an additional security buffer zone, and set in a barbed wire fence to divide it. They also sandbagged two bunkers for use as manned sentry posts. The bunkers couldn't be dug in because the parking lot had a macadam surface. Later a gate was put in to control vehicle access to the front of the BLT building.

Early on, the Marines at the checkpoints dug regular chest-high fighting holes, which were also sandbagged and rigged with overhead cover. "The main protection that we were trying to obtain here initially was from small-arms fire and overhead variable time fragmentation [ordnance] which might be used against the troops."[28]

Meanwhile, the Political/Military Committees for Beirut II were functioning differently from the way they did for Beirut I. During Beirut I, the Multi-National Force was assigned specific tasks, including evacuation of the PLO. Beirut II operations, on the other hand, were characterized by a lack of specific military tasking beyond that of military presence in specific operational areas. Ambassador Habib personally provided overall coordination and planning for the MNF during the first deployment. During the second, however, ". . . in his absence, no specific individual provided the same degree of overall coordination and direction. As a result, the French, Italian, and U.S. contingents of the MNF conducted operations in their respective areas in accordance with directions received from the national authorities of each nation."[29]

Chaired by the Lebanese Armed Forces G-3, the Military Committee met daily from 1100 to 1200, and was comprised of representatives of each MNF contingent and the LAF general staff. As in the first deployment, there were no Israeli Defense Force representatives on the committee, for all contact and coordination with the Israelis was conducted by the Government of Lebanon or through diplomatic channels. Actually, the Military Committee functioned as no more than a conduit for the flow of information, rather than as a central point for coordinating military activities. During the entire period of Beirut II, Lieutenant Colonel Smith, the MAU executive officer, was the MAU representative to the committee.

Little has been said so far about HMM-261 operations. Except for a CH-46 based at the airport for use in emergency medical evacuations, the entire squadron remained on board the *Guam*. In any case, not all the aircraft were needed ashore, where they would have

The reinforced concrete MAU Service Support Group headquarters building at Beirut International Airport. To the left (west) is the Beirut-Airport highway guarded by the LAF.

USMC Photo by SSgt Robert E. Kline

provided a tempting target for terrorists. In addition to providing logistical support to TFs 61/62, the Marine helicopters kept busy operating the "Cammie Cab Company," ferrying Ambassador Habib and his associates.

During the period 29 September-1 November, the squadron flew 888 hours with a helicopter availability rate of 92 percent. The Marine pilots transported 7,011 passengers and hauled 1,139,090 pounds of cargo.[30]

The initial logistical support for Beirut II was provided through the Seaborne Mobile Logistics System. Essentially, this encompassed the seabased warehousing of MAU supplies, which would be sent ashore on call. Although this system worked well for operations ashore of two weeks' duration or less, it couldn't support longer ones.[31] The MAU Service Support Group handled all MAU logistics requirements. In addition, it set up two shower units ashore, a water point to provide water, and a laundry unit.[32] The MSSG was also responsible for maintaining all MAU ground equipment, as well as for setting up a 30-bed hospital ashore if needed. Major Barnetson's command included two dental units, which were sent ashore. The MSSG was also responsible for all shore party operations; for distributing all supplies which landed over Green Beach, as well as those which arrived by air; and distribution of rations to all hands.[33]

Of all the many media "color" stories which came out of Beirut during this deployment, the ones concerning the feeding of Marines stand out. According to the reporters, the Marines were issued only C-rations as their mainstay, while the French were dining a la haute cuisine, with coq-au-vin and the like for their main courses; and the Italians had tables laden with several types of pasta, meats, and sauces. On the tables of both the French and the Italians were bottles of wine. All that appeared to be missing were candlelight and violins. A story that the news and television reporters did not file was about the high rate of dysentery suffered by the French and Italians, and the relatively low rate of gastroenteritis amongst the Marines. Major Barnetson said, "Dysentery among the Marines was higher in Naples than it was in Beirut because in Beirut we ate C-rations."[34] In addition, the 32d MAU had studied closely the lessons learned from the 1958 deployment in Lebanon. Initially, C-rations were alternated with MREs (meals-ready-for-eating),[35] a relatively new series of foil-packed rations which the Marines liked because of their menu variety and because they were less bulky than the canned C-rations to carry in the field. As operations ashore progressed, the Phibron ships provided the Marines with one hot meal a day, and soon two. In addition, good-hearted souls in the United States took pity on what they perceived as starving Marines, and sent thousands of frozen hamburgers and burritos to Lebanon. When possible, these were thawed and heated on the ships and distributed ashore.

In reviewing its activities in October 1982, the MAU reported, "Our efforts [to create an environment of stability by our presence] were successful as order was quickly restored and the confidence of the population gained to such an extent that rebuilding of homes and businesses commenced almost immediately."[36] For the first time in a number of years it was relatively peaceful in Beirut. There was some slight Marine concern about the Lebanese Army sweeps of west Beirut which invariably provoked large demonstrations by the Palestinian population north of the MAU positions, particularly at the site of a mosque just north of the airport. Marines were never directly fired upon by the Lebanese troops, but during Beirut II, the Marines received "errant" LAF small arms fire, generally activated by the Palestinian demonstrations.

Marine relations at this time with the Muslims who lived near the MAU area were noteworthy. The Muslims ". . . waved to, slapped hands with every Marine with whom they came into contact. They returned the wide smiles of our proud young peacekeepers. They cried!"[37]

During the remainder of October, the Marines continued to harden their positions and make "quality of life" improvements wherever possible. To facilitate a closer relationship with the other members of the MNF, 32d MAU Marines participated in several sporting events with them.

There was considerable American and international press interest in the Marine deployment, as evidenced by the plethora of stories filed from Beirut by the writing press and hundreds of videotape cassettes sent to home networks by the television media. In addition, many high officials and senior officers visited the Marines. On 12 October, Deputy Secretary of Defense Frank Carlucci and his party arrived at the airport and toured the Marine positions. The next day, the FMFLant commander, Lieutenant General John H. Miller, visited the MAU, and was briefed by Colonel Mead and his staff. Later the FMFLant staff members accompanying General Miller met with their MAU counterparts to discuss matters of mutual interest. Congressman Charles Wilson of Texas arrived on 17 October for a tour, while Vice Admiral Ronald J. Hays, CinCUSNavEur, and Sixth Fleet commander Vice Admiral Rowden visited TFs 61/62 on the 20th, when they

Commandant of the Marine Corps Gen Robert H. Barrow takes the salute of Nashville *(LPD 13) sideboys during visit to 32d Marine Amphibious Unit and Phibron 4, late 1982.*

met with the Phibron and MAU commanders both on the *Guam* and ashore. At various times throughout October, the Marines were visited by American embassy and Lebanese government officials, all of whom were given a Cook's tour of the Marine positions and headquarters.

Preparations for the relief in place of the 32d MAU by the 24th MAU began on 17 October with the arrival of a small liaison team headed by Colonel Thomas M. Stokes, Jr., the 24th MAU commander. A relief in place would remove both MAUs from planned NATO exercises, but the 32d had already been extended beyond its planned rotation date, and it was due to go home. Nine days later, on 26 October, the 24th MAU advance party arrived in Beirut to work with the 32d MAU staff on the relief plan. As the 32d began loading its equipment on Phibron 4 ships, key members of the 24th landed and made a reconnaissance of the positions ashore with unit commanders they were to relieve.[38]

Beginning at 0830 on the 30th, 24th MAU Marines moved ashore to take their assigned positions. The relief in place was completed nearly five hours later. The 32d completed reloading its equipment and personnel on board Phibron 4 shipping by 2300. At midnight on 1 November, Colonel Stokes relieved Colonel Mead as Commander, Task Force 62, whereupon Phibron 4 shipping steamed from Beirut, heading for Al Hoceima, Morocco, and an amphibious landing exercise on 9 November. The Marines backloaded from Morocco on the afternoon and evening of 10 November and celebrated the 207th Marine Corps Birthday in traditional fashion, complete with cake-cutting ceremonies. The MARG set a course for Rota, Spain, arriving there on the 14th, where the MAU landed and washed down all wheeled vehicles, heavy equipment, tanks, and AAVs. This washdown was required by U.S. Department of Agriculture regulations in order to remove Mediterranean snails from all Stateside-bound equipment. Several years before, a snail infestation of North Carolina had resulted in massive crop damage.

Two days later, the MARG headed west for the United States. While underway, Marines and sailors prepared for unloading at Morehead City upon arrival on 24 November. A Virginia State Highway representative boarded ship at Rota, and while crossing the Atlantic, lectured extensively and distributed literature about safe driving. At the same time, the MAU chaplains held classes for all hands to prepare for homecoming and reunion with their families.[39] After all the official and media attention they had received in Beirut, the Marines of the 32d MAU had become fairly blase about the press coverage of their activities. They were therefore unprepared—especially the Vietnam veterans who remembered their less-than-open-arms welcome home when they returned to the United States—for the tumultuous reception they received when they docked at Morehead City on 24 November. Captain McCabe, Company E commander, was on the hangar deck of the *Guam*, when one of his young Marines came up to him and said, "'Hey, sir, you won't believe it, go look!' and I stuck my head out [and there were] two bands, these cheerleaders, girls, people all over the place, waving. It was moving. I'm getting chills just thinking about it now."[40] In addition to bands at dockside, including the 2d Marine Division Band, playing "Semper Fidelis" and "The Marines Hymn," there were senior 2d Marine Division and 2d Marine Aircraft Wing officers, led by the division commander, Major General Alfred M. Gray, Jr. There were also throngs of people, network and local television crews, and some family members. Cheering and waving their hands, people were lined up along the route all the way back to Jacksonville. There were signs reading, "Welcome Back, 32d MAU. Good Job. Welcome Home Marines." Said McCabe, "One lady ran out of the hairdresser's with her hair in curlers, with the bib still on her, waving at us. And then her hairdresser walked out and started waving."[41] The nation's televisions screens that evening before Thanksgiving 1982 were filled with moving scenes of a heartfelt "welcome home" to the Marines for a job well done. Not since the return of the Tehran hostages—nearly two years earlier—had there been such an outpouring of patriotic fervor.

On the 24th, HMM-261 launched from the *Guam* for an official welcome as its helicopters touched down at Marine Corps Air Station, New River. The MAU headquarters, the MSSG, and the BLT were met by General Gray and their families as their buses pulled into Camp Geiger, home of the 8th Marines. To Lieutenant Colonel Blankenship, ". . . it was very uplifting. I think the welcome home helped all of us. Well, it did me, anyway, because I remember coming home from Vietnam three times . . . certainly to the young Marines it was just tremendous."[42]

Official plaudits had come from the Commandant earlier in the month, after the MAU had left Beirut. He sent the following personal messages to Colonels Mead and Stokes:

A causeway brings 24th Marine Amphibious Unit equipment ashore from Amphibious Squadron 6 ships off Beirut during the relief of the 32d MAU in November 1982.

USMC Photo

For Colonel Mead: Please convey the following message to all 32d MAU Marines.

You have successfully completed your second deployment into Lebanon, acquitting yourselves honorably and with notable distinction. Your participation in the multinational force has brought stability to Beirut for the first time in 7 years and has gvien the Lebanese citizens the opportunity to begin rebuilding their city and to commence a return to a normal lifestyle. More importantly, your efforts in stabilizing Beirut have allowed the government of Lebanon to reestablish its authority, hold a presidential election and convene the National Assembly, all key to a stable and peaceful Lebanon. Your outstanding representation of our country and the successful execution of an extremely sensitive and difficult task have added another bright chapter to the history of our Corps. Your professional accomplishments will also enhance our 207th Birthday Celebration. I wish each of you success in your upcoming Phiblex [amphibious landing exercise] and Godspeed in your return home.

For Colonel Stokes: Please convey the following message to all 24th MAU Marines:

You have assumed the watch in a clearly dynamic and changing situation that involves the maintenance of a mission that is difficult and extremely important for peace in Lebanon and the entire region. You will be required to maintain the momentum of your predecessors with expanded responsibilities and challenges. Restraint and discipline will be of the utmost importance in your very delicate situation. Diplomatic initiatives currently underway depend on the security and presence you provide in Beirut. Your superb ability to move rapidly from participation on the northern flank in Bold Guard 82 to the southern flank for Display Determination 82 epitomized your professionalism. Beyond that, speaking for myself and your fellow Marines, rest assured we have every confidence that you will superbly represent the country and Corps as a member of the Multinational Force.[43]

By 15 November, the 24th MAU had been two weeks into its deployment, while the 32d MAU, back at Camp Geiger, was getting ready with its newly joined elements, Battalion Landing Team 2/6, HMM-264, and MSSG 22, for a return to Beirut in early 1983. Meanwhile, the 32d MAU staff looked foreward to Christmas at home, while the 24th MAU got ready for Christmas in the field, not many miles away from where it all began.

CHAPTER 4

Beirut III—An Expanded Experience 1 November 1982-15 February 1983

Since 1948 in the post-World War II era, Marines have been in the Mediterranean in at least battalion strength. Then, during a crisis in Greece, President Harry S. Truman ordered the 8th Marines, at one-battalion strength, to join the Sixth Fleet as its landing force. In recent years, with the establishment of the Marine Air-Ground Task Force (MAGTF) concept, Marine Amphibious Units—each comprised of a MAU headquarters, a battalion landing team (BLT) [reinforced infantry battalion], a composite helicopter squadron, and a service support group—have acted as the Sixth Fleet's permanent landing force. The presence of the MAU in the Mediterranean gave it an opportunity to conduct amphibious landing exercises with similar units representing NATO allies situated on the Mediterranean littoral. In addition, the MAUs participated in extensive NATO exercises in the northern tier, e.g., in Norway and Denmark, where Marines figure in NATO contingency plans.

The MAUs were generally deployed for six months at a time. While squadron and BLTs would change for each deployment, with the rotation of units, however, it was conceivable that a MAU could begin its second and succeeding deployment, or "pump" as they were called, within a year and a half of its last one. When Colonel Stokes' 24th MAU left Morehead City on 24 August 1982, 40 to 50 percent of his Marines were veterans of earlier Mediterranean deployments.

Before the BLTs and helicopter squadrons joined the MAU and then went aboard their assigned Phibron shipping for pre-deployment exercises, they had already spent approximately six months training separately at first and then together as a MAGTF. When a MAU finally departed for its semi-annual deployment, its units had been tested and were certified to be fully capable of conducting amphibious operations and other tasks they might encounter during their Mediterranean duty. By the time a BLT and a helicopter squadron returned from a six-month deployment, during which time they had conducted at least one amphibious landing exercise per month with forces of other nations, they rated among the best-trained and most combat-ready units of their kind in the Marine Corps.

Colonel Stokes' 24th MAU was comprised of BLT 3/8 (Lieutenant Colonel John B. "Black Jack" Matthews), HMM-263 (Lieutenant Colonel William H. Barnes, Jr.), and MSSG 24 (Major David N. Buckner). Its strength was 1,929 Marines and 108 Navy. When the MAU arrived in Beirut, it was reinforced by 183 augmentees who had been previously attached to the 32d MAU.

The 24th MAU left the States on schedule. Since it had returned from its previous January-June 1982 deployment on 29 June, it experienced a short turnaround. The Marines were embarked on the ships of Amphibious Squadron 6, Commodore Vernon C. Smith (Captain, USN) commanding. The Phibron consisted of its flagship, the *Inchon* (LPH 12), the *Shreveport* (LPD 12), the *Fort Snelling* (LSD 30), the *La Moure County* (LST 1194), and the *Sumter* (LST 1181). The flotilla headed directly for its commitment, participation in NATO Exercises Northern Wedding and Bold Guard (East) in Scandinavia.

After the landings in the north, the MAU was scheduled to make port visits in Ireland, Holland, and Portugal before entering the Mediterranean to relieve the 32d MAU. However, as the 24th left Lolland, Denmark, it received word that these port visits were cancelled and that it would replace the 32d in Exercise Display Determination, a major NATO exercise on the southern flank of the NATO countries.[1] The MARG steamed south for the Mediterranean, where it came under the operational control of the Sixth Fleet on 30 September.[2]

Colonel Stokes was not unacquainted with the Lebanon situation, for all during his January-June 1982 float, his 34th MAU had been continually updating its intelligence and general information on Beirut, ready to undertake one of two, at that time plausible, missions: either ". . . to evacuate U.S. nationals in the embassy or reinforce the embassy."[3]

During the trip to Denmark and then later to Beirut, the BLT and squadron continuously conducted training when they were not involved in the landing exercises. The MAU intelligence section monitored incoming intelligence summaries and reports, and in turn conducted counterintelligence briefings that generally concentrated on terrorist activities. In addition, MAU S-2 personnel briefed the individual rifle companies and the MAU, BLT, MSSG, and squadron headquarters staffs on the situation in Lebanon, and

USMC Photo

An exploded car-bomb littering the beach highway below Beirut unceremoniously greeted the arrival in Lebanon of the 24th Marine Amphibious Unit in November 1982.

screened MAU personnel to identify French- and Italian-speaking Marines to be used as interpreters. With the eventual landing in Beirut in mind, each of the MAU commanders checked to see that their troops and equipment were ready.

On 11 October, the MAU/Phibron completed re-embarkation from Saros Bay, Turkey, after completion of Exercise Display Determination 82, and headed for a port visit to Naples. Concurrently, key 24th MAU staff and command personnel made a liaison visit to Beirut. An advance party flew from Naples to Beirut on 26 October, the same day that the Phibron left, to prepare for the relief of the 32d MAU. At this point, Colonel Stokes briefed his officers and staff noncommissioned officers in depth about the politico-military factions in Lebanon, rules of engagement, and standards of conduct for Marines in Beirut. Then the remainder of the Marines received a similar briefing.

As noted earlier, the relief took place in Beirut on 1 November. Under normal conditions, MAU reliefs in the Mediterranean were conducted at Rota, Spain. These were not tactical reliefs, but merely staff and organizational briefings, where ". . . you turn over certain contingency packages, maps, and Sixth Fleet plans, and then I salute Admiral Rowden and say, 'I'm ready to take over,' and Jim Mead says, 'I'm prepared to be relieved,' and then he would sail on or vice versa."[4]

Since this was not to be the case this time, Colonel Stokes prepared a landing plan wherein he would relieve the 32d's three line companies in helicopter-borne and over-the-beach landings and then he:

> . . . placed the requirement to have a mortar and anti-tank capability on the beach before—and my communications with the Navy and with the adjacent multinational force commands—I required those nets to be set up and that force to be on the ground before I said, 'I'm ready to relieve you as Commander, U.S. Forces Ashore, Lebanon.'[5]

The relief went quite smoothly and all elements of the 24th MAU quickly established a firm work routine ashore. At 1115 this first day in Lebanon, the reality of Beirut was brought to the newly arrived Marines when an automobile car bomb exploded in the vicinity of the entrance to Green Beach. No faction claimed credit for the explosion nor was it ever discovered who did it.

On the 2d, Assistant Secretary of Defense Francis J. West, Jr., a former Marine, visited the 24th MAU headquarters to become the first of the VIPs to take the Beirut tour subsequent to the departure of the 32d

MAU. He was followed on the 5th by Congressmen John P. Murtha, Robert L. Livingston, and Nick J. Rohall. It soon became apparent to this and successive MAUs in Lebanon that a protocol officer would have to be appointed to meet the important visitors and guide them around, and that a command briefing would have to be developed. Vice Admiral Rowden also visited the Marines on 5 November. Two days later, Brigadier General Andrew W. Cooley, JCS representative to Ambassador Draper, visited the MAU. On 10 November, the 207th Marine Corps Birthday was celebrated in the field with traditional Marine Corps spirit; Ambassador Dillon was the honored guest. "Thru the combined effort of BLT 3/8 and USS *Inchon* food services personnel, over 1,400 sailors and Marines enjoyed an appropriate feast with which to celebrate the birthday."[6]

Members of the staffs of Senators Paul Laxalt and Howard H. Baker, Jr., visited the Marines on 20 November, while three days later, U.S. Ambassador to Cyprus Raymond C. Ewing arrived at the MAU headquarters. On 25 November, Colonel Stokes and

Photo courtesy of Col Thomas M. Stokes, Jr., USMC (Ret)

Col Thomas M. Stokes, Jr., 24th MAU commander, is seen with Col Elias Khalil, Lebanese liaison officer.

his staff hosted JCS Chairman General John W. Vessey, Jr., who was occompanied by Admiral Rowden.

The MAU's mission was expanded on 1 November, when the Secretary of Defense approved the conduct of daylight motorized patrols, first to east Beirut and then in Baabda, and foot patrols later in Yarze. These

Seemingly oblivious of the American presence, Lebanese civilians go about their business as Marines of the 1st Squad, 1st Platoon, Company K, BLT 3/8, conduct their first patrol in Hay es Salaam ("Hooterville"), outside of the Beirut Airport in December 1982.

USMC Photo by Sgt Christopher Grey

USMC Photo

HMM-263 headquarters ashore at LZ Rockpile at the north end of Beirut Airport.

patrols began on 4 November after extensive planning and conversations between the U.S., French and Italian MNF officials, and Lebanese military authorities, who helped establish routes and timing of the patrols.[7] The first patrol consisted of four jeeps with machine guns mounted. Two jeeps went out first, followed five minutes later by two more. The patrol had ". . . 160s [radio set AN/GRC-160], thereby having the ability to talk inter-patrol on the VHF [net] and we used an HF net to talk back to the BLT and the MAU."[8] A total of 15 Marines and a Lebanese liaison officer, acting as an interpreter, went on the patrol. Continuous communications between the patrol and the MAU headquarters were facilitated by the use of an automatic retransmitting station, airborne in a HMM-263 helicopter with a back-up manual retransmitting station at the Presidential Palace. The first patrol went out for two and a half hours on the afternoon of the 4th, the second went out for two hours the next morning. Both returned without incident.

Thereafter, patrols went out daily. By the end of November, 30 patrols had been successfully conducted. With JCS approval, the patrol routes had been expanded to cover northeast Beirut. Aside from the military aspects of these patrols, there was another dividend, and that was one of giving the Marines a feeling that they were doing something historic, that ". . . they contributed . . . to the stability of the Beirut area and . . . to world history as Marines."[9]

During November, the 24th's intelligence section was augmented with the arrival of detachments from the Sensor Control and Management Platoon (SCAMP) and interrogator/translators from the 2d Marine Division at Camp Lejeune. This combination of human intelligence (HUMINT) and sensor assets was employed to provide adjacent, subordinate, and higher commands with a good variety of intelligence information. Colonel Stokes noted that ". . . this full-scale intelligence collection and dissemination program has satisfied the commander's essential elements of information and other concerns in the 24th MAU area of responsibility."[10]

The character and nature of the 24th MAU's tour in Lebanon was different from the 32d's—as a matter of fact, the tours of each of the MAUs differed from the others considerably, usually in four areas: weather; training emphasis; relationship with other forces in the area; and finally, the nature of the fighting in the surrounding area. The weather which had been at first warm, then hot, when the 32d's Marines arrived, had become pleasantly fall-like when the 24th first came in and then became cold and wet in the succeeding winter months. The surf at Green (Black) Beach got quite heavy during the winter, and seriously disrupted over-the-beach supply operations. HMM-263 helicopters took up the slack by flying in needed supplies from shipboard to the landing zones at the beach and in the MAU perimeter. Despite the weather, however, the helicopters were able to fly in two hot meals a day from Phibron shipping.[11]

At first, the MAU had begun a limited training program with the Lebanese Armed Forces (LAF). Then on 11 November, in response to a request from the Government of Lebanon, the Joint Chiefs of Staff proposed that the Marines begin training a Lebanese Army rapid reaction force. The training would be conducted only if it did not interfere with the Marines' basic mission. On the 12th, the MAU provided a training plan for approval by higher echelons, gaining that approval on the 30th. Ground units and air crews to be trained were designated by the Lebanese government. Training began on 13 December.

The initial training was conducted in three one-week phases. Phase I consisted of training in general military skills and physical training, and an orientation briefing on Marine Corps weapons and equipment. The second phase consisted of helicopter orientation, live firing exercises, an overview of amphibious operations, a visit to Phibron 6 amphibious ships, and instruction in antimechanized operations. Phase III training consisted of planning for helicopter-lifted assaults and extractions. Paralleling this instruction was training given to Lebanese air crews in all aspects of helicopter assault support. Phase III ended on 7 January with a demonstration of a vertical assault operation.

A 24th MAU Marine trains Lebanese Armed Forces soldiers in close combat techniques in December 1982.

Photo courtesy of Claude Salhani

Some Lebanese officers had attended Army schools at Fort Benning and Fort Leavenworth, and although fairly well-versed in general military subjects, they were rusty in military skills. The most serious weakness in the Lebanese Armed Forces, however, was the inexperience of their non-commissioned officers.[12] Every senior Marine instructor was a gunnery sergeant, and as Lieutenant Colonel Matthews, BLT 3/8 commander, recalls:

> . . . our goal was to allow the Lebanese Army to see how our NCOs function, and they function without officers and they saw that and . . . in many cases absorbed that kind of demonstrated leadership, and they certainly absorbed a lot of our spirit.[13]

Lieutenant Colonel Matthews also noted that his Marines trained battalion after battalion of the Lebanese Armed Forces in close combat and bayonet training, ". . . and they hear the arrrugahs going, they seemed to enjoy it."[14]

To demonstrate the seriousness of its intent in rebuilding the LAF, on 11 December, the Government of Lebanon appointed as commanding general of Lebanese Armed Forces, General Ibrahim Tannous, a barrel-chested, war-scarred veteran. He seemed determined to rebuild an organization that really had not been out of its barracks since 1976 except to man permissive checkpoints. Lieutenant Colonel Matthews saw General Tannous as a carbon copy of his division commander, tobacco-chewing Major General Al Gray.

Lieutenant Colonel Matthews noted that Tannous':

> . . . rapport with the troops is almost the same as General Gray's . . . General Gray's got the capability of talking with the PFCs and they know he really cares about them. And that's the way General Tannous comes across, in my view, with the Lebanese Army. So, I think that's awfully good and healthy.[15]

Both the Marines and the Lebanese profited from the cross-training program. For the Lebanese, it meant a shaping-up of basic skills, if not, in fact, learning them for the first time. For the young Marines it meant sharpening their own skills, ". . . and working with the Lebanese soldiers gave our troops a very clear mission identification. The Lebanese whom they knew . . . they worked with, squad leader to squad leader, . . . did have a real strong desire to defend their country."[16]

USMC Photo

A BLT 3/8 platoon leader demonstrates the TOW anti-tank missile launcher to members of the French MNF.

During January, in preparation for artillery cross-training with Lebanese artillery units, Battery G, 10th Marines, the 24th MAU's artillery unit that came ashore on 3 December, began training in all aspects of its specialty. Cross training with the LAF began on 10 January. There was also cross-training in combat service support, for the Lebanese especially needed familiarization with the new vehicles and equipment the United States was supplying their army.

Cross-training also began with other MNF units in December. That month, the MAU's amtracs began using the Italian force's tracked vehicle course to maintain the proficiency of Marine drivers and mechanics. On the 14th, the Marines conducted a training exercise with 102 French paratroopers. The exercise included a heliborne assault demonstration, amtrac operations from the *Fort Snelling*, and weapons familiarization ashore. Later that month, the French paratroopers joined the Marines in a training session that involved rappelling from helicopters.

As soon as the 24th MAU's troops were settled in, their own unit training began. Much time was spent learning about the culture and history of Lebanon. The political officer of the American embassy presented lectures to the Marine officers on the Lebanese political situation. Professors from Beirut University, from the American University in Beirut, and from the faculty of the Lebanese Science University talked to the officers, who in turn ". . . would impart that information to the troops and we'd go down and talk to the PFC and he's telling you about the Druze and the Phalange; you know, he had a crash course of 109 days in the history of Beirut proper and Lebanon in general."[17] The dividend from all this was the fact that the young Marine could go home after his tour in Lebanon and speak intelligently about his experiences.

Tours of historic Lebanese places, begun during the 32d's tour, were continued by the 24th. Although none of the Marines was allowed liberty in Beirut, Colonel Mead had introduced a program of getting his Marines out of the lines and into trucks for tours of Beirut, past the famous Museum Crossing which separated Muslim west Beirut from Christian east Beirut, and Martyr Square, and then to Juniyah. While the 24th MAU was in country, representatives of the Lebanese Ministry of Tourism also took Marines on tour of Byblos, an historic coastal village about 20 miles north of Beirut. Called Jubayl in olden times, it dated back to the days of ancient Phoenicia and contained historic ruins, remnants of its earlier splendor. The Marines also were able to go on one-day skiing excursions into the mountains.[18] Sailors from Phibron 6 and the carrier battle group were also invited to join in these trips. The situation ashore at this time also permitted the MAU to send some of its Marines on port visits with Phibron ships to Athens, Greece, and Antalya, Turkey.

While the Marines were performing their daily tasks on the ground, and the squadron's helicopters were busy with either logistic or diplomatic support missions, the MAU's doctors and corpsmen established a Medical Community Aid Program (MEDCAP) on 30 November, with the assistance of the Lebanese liaison officer to the MAU. In an area on the airport road north of the MAU compound, at a Lebanese checkpoint, the MSSG set up three general purpose tents on a hard-top site, 100 meters by 100 meters, and wired the tents for lighting. Three days each week, the MAU's medical platoon trucked medical supplies and dental equipment to the tents, and prepared to open sick call at 0900. Treatment was given to all Lebanese who sought it. On the medical end, the patients were diagnosed and treated, but no surgery was performed. The dental section, on the other hand, ". . . was able to provide a level of dentistry that far exceeded anything that even the most affluent Lebanese could purchase in Beirut."[19] Initially, the Lebanese were hesitant to accept the services the Americans were providing, but before long, there were lines of patients waiting for treatment. Before the 24th MAU left Beirut, its MEDCAP had treated over 2,000 Lebanese nationals.[20]

During December, Colonel Stokes and his command were visited by General Tannous and President Gemayel, as well as by Major General Gray, and Terence Cardinal Cook, Military Vicar of the U.S. Armed Forces, who celebrated mass while ashore. Four British liaison officers visited Colonel Stokes in January, prior to the arrival of a British contingent for the MNF in February. Congressman Murtha revisited Beirut in January. Before the MAU left Lebanon in February, it also hosted two separate senatorial groups, one headed by Senator Dan Quayle and the second by Senator Dennis DeConcini, as well as permanent staff members of the Senate Foreign Relations Committee. Military visitors to Task Forces 61 and 62 included Vice Admiral Thomas J. Kilcline, Commander, Naval Air Forces, Atlantic Fleet; the Inspector General of the French MNF contingent; Vice Admiral Charles R. Larson, Commander Task Force 60, *Nimitz* Carrier Battle Group; and Vice Admiral Edward S. Briggs, Commander, Naval Surface Forces, Atlantic Fleet.

Photo courtesy Col Thomas M. Stokes, Jr., USMC (Ret)

24th MAU commander Col Thomas M. Stokes, Jr., escorts President of Lebanon Amin Gemayel (right) and Gen Ibrahim Tannous (left), Lebanese Armed Forces commander, as they make a Christmas Day visit to the 24th MAU in 1982. To the left of Gen Tannous is 24th MAU Sergeant Major Valdemar Vasquez.

Terence Cardinal Cooke, Military Vicar of the Armed Forces of the United States, visits Marines and sailors of the 24th Marine Amphibious Unit, Beirut, New Year's Eve, 1982.

Photo courtesy of Claude Salhani

USMC Photo by SSgt Robert E. Kline

2d Marine Division commander MajGen Alfred M. Gray visits with Lebanese soldiers.

A matter of concern occurring during this deployment was the escalation of Marine-Israeli Defense Force confrontations. To Colonel Stokes, it was a real problem. He was not only worried about the possible loss of life, but also about the impact on Israeli-American confrontation would have on the Lebanese situation overall.[21]

The problem stemmed from the Israelis being assigned the Old Sidon Road to use as a main supply route for their troops in positions northeast of the Marines. In the latter part of November and through all of December, Israeli vehicles were being fired upon. The IDF then dispatched convoys with armed escorts. At the end of December, a high ranking Israeli officer was reportedly killed in an ambush. In early January 1983, the command vehicle of a convoy was destroyed with loss of life. The Israelis suspected that the attacks were being mounted by PLO personnel who emerged from Marine-occupied territory and then immediately retreated to what they considered sanctuary.

The first Marine-Israeli contact occurred on 5 January, when an Israeli tank entered Company K's positions in the eastern portion of the Marine perimeter. Claiming to be lost, the Israelis were quickly escorted out of the Marines' territory. Colonel Stokes happened to be visiting Company K that day, and he ". . . refreshed the [Israeli] tank company commander's memory on the extent of USMC boundaries around the Beirut International Airport and [on] land navigation."[22]

Again, on 6, 8, and 10 January, the Israelis attempted to enter U.S. positions and to set up direct conferences between Colonel Stokes and their commander. In each case, the Israelis were not allowed into Marine lines, and they were reminded that requests to confer with Colonel Stokes had to go through diplomatic channels. In commenting on this matter of dealing with the Israelis on a face-to-face basis, Colonel Stokes later expressed some of his frustrations and a military professional's point of view by saying:

> Ground commanders do a much better job of dealing with and clarifying their own tactical matters than do staff officers and diplomats. If there had been a free and timely flow of required info between this officer and Brig Amnon [Lifkin, senior Israeli officer in the area] and LtCol Matthews [CO, BLT 3/8] and LtCol Landsberg (Israeli tank unit commander], the last 30 days may have been much quieter for us in south Beirut. The above may not be possible, but it is logical.[23]

In addition to the incursions into their territory, Marines had to contend with the repeated Israeli patrol practice of reconnaissance by fire, which they began on 9 January. "These patrols were characterized by intermittent firing of small arms, main tank guns (firing usually being directed toward vacant buildings and into open fields or tree groves both west and east of the Sidon Road)."[24] The Sidon Road patrols generally moved from south to north, and upon reaching the proximate position of Marine Company L, the patrol would move south and take up a position, where it remained all day. These patrols were generally comprised of one to four armored vehicles (armored personnel carriers or tanks), followed closely by 5 to as many as 14 dismounted soldiers. Although the patrols became predictable and routine, the firing clearly became a threat to the safety of U.S. forces. It was quite clear to Colonel Stokes that he had to discuss the matter face to face with the senior Israeli officer in the area, and so he insisted upon such a meeting. With the assistance of Ambassador Dillon, he was able to arrange one.[25]

Photo courtesy of Claude Salhani

LCpl Roger E. Dery (l) and Cpls James Burnett (center) and Darrell A. Banks (r) contemplate the distances from Beirut to some places they would rather be.

Colonel Stokes met with Brigadier General Lifkin to discuss the Sidon Road problem and other matters of mutual concern. The Israeli general agreed to establish a direct radio link between his headquarters

Copyright 1983. Don Wright. *The Miami News*. Reprinted with permission.

and that of the Marines. This was done on 30 January. Four days earlier, the Israeli practice of reconnaissance by fire had ended, although patrolling continued.[26]

In February, the understanding with the Israelis over boundaries and the conduct of patrols—which was thought to be a settled matter—was found to be not so clearly understood as originally thought. The single-most notable demonstration of this lack of understanding occurred on 2 February, when three Israeli tanks attempted to go through Captain Charles B. Johnson's Company L position.

At about 0800, from his observation post, Captain Johnson, together with the advance party of the British MNF contingent,[27] observed an Israeli patrol coming up Old Sidon Road from the south. This was normal. Half an hour later, he spotted a north-to-south patrol, which also was normal. It consisted of three tanks, two armored personnel carriers (APCs), and dismounted troops. "Again, we're seeing them about 3,000 meters off. We could see that far, all the way down the Sidon Road."[28]

The only thing that was unusual about this patrol was that the troops were dismounted, for the Israeli patrols in the previous two weeks had all been mounted. Captain Johnson then went on to say:

> . . . sometime between 0830 and 0900, one of my surveillance people . . . spotted three additional tanks coming on the road . . . the one they had built along the railroad tracks, and then they [the tanks] broke off the road and they continued up the railroad tracks right up to the edge of the university grounds. . . . That's when I knew somthing was up. There were three tanks on the road. . . . There was no tactical reason for them to do that. . . . They brought tanks right through the middle of Shuwayfat, which is a Muslim area and it's relatively dangerous to do that.[29]

What Captain Johnson had spotted were three tanks coming from the north and three tanks coming from the south. He couldn't see them when they were in the town, but they were spotted shortly after as they left it and broke through the orchard on the western side of the Sidon Road into the buffer zone between the road and the university. The tanks were heading for a section of the fence where Captain Johnson had confronted an APC-mounted Israeli patrol on 20 January. The Company L commander quickly got in his jeep and went to the spot the tanks were approaching. Captain Johnson didn't think that:

> . . . they would actually try to come through a joint Marine-Lebanese checkpoint like that. But once it developed, I was very concerned that if the tanks were allowed to move forward, there was a very dangerous situation, because the road they were on . . . went right through the heart of the

university . . . divided the Marine company and the Lebanese company.[30]

Johnson feared that if the tanks attempted to pass, a firefight might erupt between the Lebanese and the Israelis. If a fight ensued, the Marines would have to support the Lebanese. He wasn't worried about the Marines' fire discipline, but he was concerned about that of the Lebanese soldiers.

As the Israeli tanks approached the fence, Captain Johnson jumped out of his jeep, ran up to the tanks, and stood in the center of the road. The lead tank stopped about six inches in front of Johnson, who told the Israeli lieutenant colonel in the lead tank, "You will not pass through this position." After a short pause, the Israeli dismounted, spoke with Johnson, and then climbed back aboard the tank, saying that he was going through. Johnson later stated that he replied, "You will have to kill me first."[31] He drew his pistol, chambered a round, and held the weapon at the ready position. There was another pause as the Israeli officer apparently spoke over his radio to his headquarters. The lead tank then pulled slowly to the side of the road with Captain Johnson walking alongside and then the two others suddenly revved up their engines and whipped forward toward the fence.

The young Marine captain jumped on the lead tank, grabbed the Israeli officer, and yelled at him to order his tanks halted. The tank commander complied and then purportedly told Johnson, "One thing we don't want to do is kill each other." Johnson answered, "Yes, but if you keep doing things like this, the likelihood is going to occur."[32]

While the local Arab radio stations were telling and retelling the story of the American who stopped the three Israeli tanks singlehandedly, the Israeli press was accusing Captain Johnson of having liquor on his breath and being drunk. Worse, they called the whole affair a misunderstanding on the part of the Marines. Confronted by evidence, among other things, that Johnson was a teetotaler, the Israelis quickly toned down, and finally stopped such comments when they saw they were not going to be given credence.

Within a few minutes of the confrontation, Johnson's battalion commander, Lieutenant Colonel Matthews, arrived on the scene. He had observed part of what happened and asked Johnson for a full and immediate report, "And I gave him the whole thing . . . and we spent about 20 minutes walking the ground and so forth."[33] Matthews then said they should tell the whole story to Colonel Stokes, who went back to the fence area with Johnson and rewalked the area where the confrontation took place. The MAU commander reported the incident through the chain of command. The next day, 3 February, Israeli and American diplomats met in Beirut, where they agreed to mark the boundary lines more clearly so there would be no future misunderstandings.

A routine, daily press conference was held at 1600 on the afternoon of the 2d at Colonel Stokes' headquarters. The most important topic concerned a ricochet 75mm tank round that had landed in Company I's positions. Nothing was said about Captain Johnson's experience until the press stormed back into the compound at 2300 that evening, undoubtedly having been queried by their home offices why stories had not been filed on the U.S.-Israeli affair. When the reporters asked Colonel Stokes why he hadn't told them about it, he replied that no one had asked, and said further, ". . . it's not my job to determine what's newsworthy and what's not. . . ."[34]

Normally a quiet officer despite his impressive military presence, Captain Johnson was told by his CO that he was going to have to submit to the questions of the print and television reporters at a press conference, much as he disliked the prospects of such an encounter. A by-product of this instant fame was heavy mail. A large number of former Marines and retired servicemen wrote and sent messages of support. "A lot of children wrote from schools and they were really nice letters. A lot of people wrote. I got hundreds of letters." Captain Johnson also received a message from the Commandant after the 24th left Lebanon. "It was a wonderful message to my men, how he was proud of the men," Johnson said. In retrospect, Johnson never felt that what he had done was wrong. "I had no doubt in my mind that what I had done was the right thing. . . . I had regret that it happened, but I did not have any regret in what I had done."[35]

During the month of January, the MAU prepared for its scheduled relief in February. Like the previous October's turnover, it would be a relief in place. The advance party of the 22d MAU arrived in Beirut on 9 February and each member was taken in hand by his 24th MAU counterpart. Since the first relief had gone so smoothly, there was little reason to believe that the second would be otherwise. It wasn't. At 0700 on 14 February, elements of the 22d MAU started landing and BLT 3/8 was relieved in place by BLT 2/6 by 1251, MSSG 24 was relieved by MSSG 22 at 1300, and HMM-264 relieved HMM-263 of the Cobra alert mission at 1326. Colonel Mead, commander of the 22d MAU, back in Beirut for a third time, assumed control of the forces ashore at 1515. The next day, 15 February, he assumed command of the U.S. Multi-

Department of Defense (USMC) Photo DN-SN-84-00723

A Marine stands watch as other Marines jog on the Beirut International Airport perimeter road, with Burj al Barajinah, outside the fence, posing a threat in the background.

MAU Marines decorated "The Peacekeepers' Tavern," a spare but cool and safe refuge in the basement of the MSSG headquarters building at Beirut International Airport.

Photograph by the author

Photo courtesy of Claude Salhani

Senior Chief Journalist Tom Jones, USN, mans the control board on the Armed Forces Radio and Television Service broadcast module set up in the U.S. Multi-National Force compound at Beirut Airport.

National Force in Lebanon, as Phibron 6 with 24th MAU embarked was steaming towards Rota and home.[36]

The 24th MAU carried out its washdown in Rota and sailed to Morehead City, where it arrived on 8 March to be greeted by bands, the media, and families. Colonel Stokes was relieved as CO by Colonel Timothy J. Geraghty on 17 March. Four days later, BLT 3/8 and HMM-263 were relieved as elements of the 24th MAU by BLT 1/8 and HMM-162. For service in Beirut, the 24th MAU was awarded the Navy Unit Commendation; Colonel Stokes was decorated with the Legion of Merit.[37]

The 24th MAU's deployment was not as spectacular as the 32d's deployment nor were its Marines in the international spotlight as much. But the MAU sailed home with a feeling of a difficult job well done in continuing the diplomatic mission of presence handed to it by its predecessor. Like the 32d MAU before it, the 24th MAU had been well supported by FMFLant. The response in terms of equipment, personnel, and the like was, according to Colonel Stokes, ". . . almost embarrassing it was so damned good and fast."[38] The cold, rainy weather had been wreaking havoc with the field boots of the Marines. The MAU sought to requisition an overboot to keep the Marines' feet dry, ". . . and I had a couple of airplanes full of them before I could shake a stick. And . . . these same young kids whose mothers tried to put galoshes on them when they were about eight or nine years old—they wouldn't sell that pair of galoshes they put over their boots for $100."[39]

The 24th MAU maintained an active physical training program. Marine runners used the airfield perimeter road, which measured six miles all the way around, and a number of them participated in a 10 kilometer race with the French and Italians two weeks after their arrival in October, and lost.

The Armed Forces Radio and Television Service arrived at Beirut on 17 December. Now the MAU Marines could listen to radio programs featuring the latest news or the "Top Twenty" popular tunes. Later in the Marine deployments, the AFRTS would broadcast television programs, in which the Marines could see themselves featured in network news shows and also watch commercial television programs, without commercials.

Meanwhile the Marine flyers could list some real achievements. HMM-263 flew 6,349 hours in logistic, VIP, and diplomatic missions. By the time the Marine flyers returned to New River, they had all qualified as helicopter aircraft commanders.

MSSG 24 was also well employed during its 108 days ashore in Beirut. With its 16 MA-13 five-ton trucks, Major Buckner's truck platoon logged 39,000 miles, as opposed to the usual 15,000 registered in a normal six-month MAU deployment in the Mediterranean.[40] Careful preparation and attention to detail led to the successful operation of the service support group in Lebanon, and, according to Major Buckner, his Marines ". . . just did a magnificent job. Very heartwarming to see. And when they left, they left with a real sense of fulfillment, whether the guy was a truck driver or mechanic, or a shore party man or a dentist or military policeman or an air delivery guy, a communicator, admin guy. You know, he felt that he had really done something worthwhile. And I'm sure they got a lifetime of memories out of it."[41]

Colonel Matthews had much the same reaction when he asked one of his Marines what he thought of Lebanon. The Marine replied, "Sir, it was an experience. . . . It was a good one, because I feel for the first time in my life I've done something that is positive. I feel that I contributed something to a country that wants to get on its feet."[42]

With this the general consensus, the 24th MAU returned home. And within weeks after the 22d MAU began Beirut IV, the 24th started preparations for its May 1983 deployment.

CHAPTER 5

Beirut IV—Circumstances Change, 'Presence' Remains—15 February-29 May 1983

Colonel Mead had been selected for promotion to brigadier general in January 1983 and there was speculation that he would be "frocked"[1] so that he would be equal in grade to the heads of the French and Italian MNF units in Beirut. As he later explained, if he had been promoted in January, and had he kept his command, the MAU would possibly have had to be upgraded to become a Marine Amphibious Brigade (MAB), with the possible addition of another battalion landing team.*

In the minds of many, the prospect of the MAU becoming a MAB was not so far-fetched, for it was again speculated that if the Israelis pulled out of their positions and headed south, the void would have to be filled by extending the Marines to the south. That would require the MAU to be augmented by at least a BLT. As it turned out, all of this speculation was for naught. The 22d MAU remained a MAU and Colonel Mead was not promoted until June, after he returned to Camp Lejeune and relinquished command.

Like the 24th's turnaround, the 32d MAU's time between deployment was also of short duration. The 32d MAU returned to Camp Lejeune on 24 November, detached the BLT and the squadron, sent its personnel on holiday leave, joined up its new elements, and left Morehead City for Beirut on 27 January 1983.

The 32d Marine Amphibious Unit was redesignated 22d MAU on 1 December 1982[2] for its second trip to Beirut and began its predeployment training with Amphibious Squadron 2, Commodore (Captain, USN) George Bess, commanding. The MAU had already been joined by its new ground and air element, BLT 2/6, commanded by Lieutenant Colonel Donald F. Anderson, and HMM-264, Lieutenant Colonel Richard J. Kalata in command. MSSG 22 had a new commander, Major Albert E. Shively. The MAU-Phibron combination quickly melded into a solid Navy-Marine Corps team, as Colonel Mead's Marines looked forward to their Beirut deployment; 45 percent of the MAU staff had been there before. The intelligence and operations section continually monitored the Lebanon situation as well as the situation in the Caribbean, either or both of which might impact on the 22d MAU's deployment.

Particularly busy during the predeployment period was the MAU Service Support Group, which is tasked with providing all the combat service support required by a deployed MAU. While its strength is generally only 278-280 Marines, the MSSG is a uniquely diversified organization made up of varied platoons and detachments from the 2d Marine Aircraft Wing, the 2d Marine Division, and the 2d Force Service Support Group—the three major commands of Fleet Marine Force, Atlantic.

The MSSG is responsible for management of the MAU's LFORM (Landing Force Readiness Materiel) block, as well as the Operational Deployment block, which itself requires seven days to be moved from Camp Lejeune to Morehead City and then loaded on board Phibron shipping and stowed properly.[3] In the case of the November 1982 to February 1983 deployment of the 24th MAU, MSSG 24, under Major Buckner, devised a Rapid Deployment Block made up of anticipated high usage items which would be needed ashore. When the block was brought to the beach in Lebanon and warehoused, it provided 79 percent of the items needed by the 24th MAU. The ready availability of those items cut down on the number of logistic support flights needed, freeing the squadron's aircraft for other chores.[4]

To ensure self-sufficiency, the MSSG also included a maintenance platoon comprised of Marines with highly specialized talents. "You name it, the maintenance platoon is charged with fixing whatever breaks."[5]

With all Marines and equipment loaded, the Phibron steamed out of Morehead City on 27 January on board the *Guadalcanal* (LPH 7), the Phibron flagship; *Raleigh* (LPD 1); the *Pensacola* (LSD 38); the *Spartanburg County* (LST 1192); and its sister ship, the *Fairfax County* (LST 1193). The first night out, the Phibron experienced high seas and 60-knot winds, which resulted in, as Colonel Mead noted later, "Marines getting their sealegs fast."[6]

On 7 February, the 24th MAU liaison officer arrived on the *Guadalcanal* with turnover packages for Colonel Mead, while two days later, Lieutenant Colonel Ronald R. Rice, 22d MAU executive officer, led an advance

*Colonel Mead comments to author, 23May83. However, when Brigadier General Jim R. Joy became CO of the 22d MAU in November 1983, the 22d remained a MAU.

Photo courtesy of Francoise de Mulder

During a quiet period, 22d MAU Marines fill sandbags near Lebanese University.

party into Beirut. As the executive officer, Lieutenant Colonel Rice would be the 22d MAU's representative to the twice-weekly meeting of the MNF Military Committee, and at the same time, he would maintain contact with the defense attache at the embassy and the EUCOM liaison officer.[7] The relief in place of the 24th MAU began on 14 February and was completed without incident on 15 February, when Colonel Mead once again assumed full responsibility as commander of the U.S. contingent to the Multi-National Force, Beirut, and Commander Task Force 62.[8] That same day, he hosted Sixth Fleet Commander Vice Admiral Rowden and Admiral William J. Crowe, Jr., Commander in Chief, U.S. Naval Forces Europe (CinCUSNavEur) at the MAU headquarters.

The troops immediately settled into a routine of daily mobile patrols in east and west Beirut, and foot patrols in the airport sector and in Baabda. The 22d MAU resumed crosstraining LAF units where the 24th MAU had left off.[9] The Marines were drilled and redrilled on the rules of engagement, including "what-if" sessions in which all possible contingencies were hopefully covered. In addition, each Marine was issued a wallet-size card with the following rules printed in all capital letters:

Guidelines of Rules of Engagement

1. When on the post, mobile or foot patrol, keep loaded magazine in weapon, bolt closed, weapon on safe, no round in the chamber.
2. Do not chamber a round unless told to do so by a commissioned officer unless you must act in immediate self-defense where deadly force is authorized.
3. Keep ammo for crew served weapons readily available but not loaded. Weapon is on safe.
4. Call local forces to assist in self-defense effort. Notify headquarters.
5. Use only minimum degree of force to accomplish any mission.
6. Stop the use of force when it is no longer needed to accomplish the mission.
7. If you receive effective hostile fire, direct your fire at the source. If possible, use friendly snipers.
8. Respect civilian property; do not attack it unless absolutely necessary to protect friendly forces.
9. Protect innocent civilians from harm.
10. Respect and protect recognized medical agencies such as Red Cross, Red Crescent, etc.

The Marines were expected to know and understand these 10 rules as thoroughly as the 11 general orders for guard duty which they had learned in recruit training, and as they knew their names, ranks, and social security numbers.

When the 22d MAU returned to Beirut, it had the experiences of its previous deployment on which to base its needs and plans. Essentially, the Marines knew the terrain and the area they were going into. The 24th MAU had improved the positions their predecessors had dug and extensively sandbagged. Filling sandbags is one way of keeping idle young Marines from becoming bored, as well as keeping them in tiptop physical condition, and in all the months of the Marine deployment in Lebanon, sandbags beyond count were filled. One task the 22d intended to undertake, but never had time for, was to cut additional fields of fire, keeping the brush down in front of some of the positions on the eastern sector of the airport in order that Marines in their fighting holes could see who was shooting at them, and from what direction, when they came under fire.

During their time in Beirut, Colonel Stokes' 24th MAU Marines had emplaced artillery in their northern sector, the boundary between Marines and the Italians. They had also improved the sandbagging in the beach area, as well as improved and weatherproofed positions at the Lebanese University, which, in the words of one Marine, had been continually "raped" by one faction or another during the previous years of fighting. The troops in perimeter positions lived in sandbagged general purpose tents, and when the 22d MAU landed in February, it brought with it additional lumber to strongback the tents, and replacement tentage.[10]

Based on what he and BLT 2/6 operations officer Major Michael L. Rapp perceived as a changed political situation while analyzing their own mission, Major Farmer, now the 22d MAU S-3, decided new positions needed to be added to the eastern perimeter of the airport in the Hay es Salaam area, which contained a wholly Shiite village. The Marine tank park was also relocated to a site where the armor could be more quickly responsive when needed. Similarly, the amtracs were relocated and dispersed, so that they could function as personnel carriers, which they actually did sometime later when an Italian convoy was attacked and the Marines were called on for assistance.[11]

The 22d brought in only a small amount of cold weather gear for Beirut IV, because Lebanese winters are usually mild. As the cold weather intensified a short time after the landing, Major Shively, the MSSG commander, purchased commercial space heaters in the open market to heat the Marines' living quarters. Field kitchens were set up in a centralized field mess for the MAU and BLT headquarters and company-sized galleys were set up at the perimeter positions. The *Raleigh* and *Guadalcanal* provided breakfast and dinner respectively until 28 February, when the central dining facility ashore became operational.[12]

Marines take time out for physical training on top of a building at Lebanese Univerity.

Photo courtesy of Francoise de Mulder

Photo courtesy of Francoise de Mulder

A patrol leader briefs his Marines before they begin patrolling in the vicinity of the American sector.

Colonel Mead knew that he wanted to bring a larger supply maintenance block ashore in his third deployment, instead of leaving it on shipboard. Of primary consideration in this was the fact that the 22d MAU's artillery and tanks would be landed and maintained ashore.[13] Because of inclement weather, it took two weeks to unload all Phibron shipping. Within four days after the turnover, instead of a normally mild winter, Lebanon began to experience its worst weather in 40 years. The temperature dropped to the low 40s at the airport, with up to 70-knot winds. These conditions prevailed until early April and made life difficult both ashore and afloat.[14]

The political situation facing the 22d MAU in February had also changed for the worse. It appeared as though the various factions in Lebanon were choosing sides, and that the terrorist threat had increased its level slightly since Beirut II.[15] In speaking of Beirut IV, Major Farmer saw the 22d MAU's mission as 80 percent political and 20 percent military. He saw no indications that the Marines would move from their airport positions. "The only additional operational requirements that we had that I did not have the first time I was in country was an increased patrolling effort and the training of the Lebanese Armed Forces."[16]

LAF training was coordinated by the MAU headquarters, but was actually conducted by BLT, squadron, and MSSG Marines. The training syllabus

A HMM-264 CH-46 helicopter lands in the mountain snow to evacuate survivors of the severe February 1983 storm which hit Lebanon, disrupting almost all transportation.

continued to stress the basic military skills the 24th MAU had taught. In addition, Marines trained the Lebanese in diesel mechanics, but trained Lebanese tankers in basic armor techniques only, since the LAF had no tanks yet. Later, a U.S. Army Mobile Training Team arrived and relieved the MAU of some of its training chores.

Conducting patrols in a potentially hostile environment also had training benefits for the Marines. Ordered to conduct a patrol within a certain area, a patrol leader would select routes in conjunction with a Lebanese liaison officer. (Over time, patrol routes and times would be varied to keep terrorists from detecting patterns.) The Marine patrol leader would next go through the doctrinal troop leading steps, organizing and conducting the patrol, and would give a thorough report afterwards. Initially, the patrols of the 22d MAU were conducted without incident.

On 20 February, before the MAU could settle into a regular routine, a heavy snowfall in the mountain area of Lebanon isolated a number of villages and stranded many travellers. President Gemayel requested MNF assistance in rescuing approximately 200 Lebanese trapped near Dahr al Baydar, about 20 kilometers east of Beirut. The next day, the MAU committed both helicopter and motorized/mechanized equipment to the rescue efforts, but Lebanese officials had to obtain Syrian clearance before the MAU could move, because the afflicted areas were behind Syrian lines.[17]

When Syrian approval came through, two HMM-264 helicopters attempted to fly to the rescue site but were forced to turn back because of heavy icing conditions. Meanwhile, the Marines' amtracs were staged at the Lebanese Ministry of Defense building waiting for the Syrians to clear an American surface rescue attempt towards Dahr al Baydar. At the same time, General Tannous told the MNF that its assistance was needed in the mountains approximately 40 kilometers northeast of Beirut, where there were no Syrians.

At 0445 on 22 February, a column of nine amtracs headed for Dahr al Baydar to join Lebanese Red Cross personnel. The ground rescuers first set out for Qartaba, while two Marine helicopters flew directly to Dahr al Baydar. One UH-1N reaching a mile-high landing site, found itself in a precarious situation. Just prior to touching down, the cockpit instruments warned that a fire control radar was locked on the Huey. Despite this threat, the pilot landed and then the lock disappeared. Five poorly dressed Syrian soldiers came up to the Marine helicopter, offered the pilot and co-pilot some coffee, and told them that a few Lebanese in stranded vehicles were located not more than a kilometer away. The first car the pilots checked had been there for three days in high winds and subfreezing weather. Its two passengers were dead. Lieutenant Colonel Kalata, the squadron commander, directed the second helicopter, a CH-46, to land in another very difficult landing zone to evacuate four survivors. Both aircraft returned to the Ministry of Defense to unload the evacuees. After refueling, they headed to Qartaba to coordinate rescue efforts with the mechanized column still en route.[18]

Photo courtesy of Claude Salhani

Following the winter storm in February 1983—one of the worst in the history of Lebanon—22d MAU LVTP-7 assault amphibious vehicles (AAVs) were employed in the evacuation of snowbound civilians.

At the same time, Italian and French columns were attempting similar rescues. The Italians started moving along the Damascus highway, but were stopped by the Syrians 10 kilometers short of their desintation, perhaps for political reasons or perhaps because the Syrians realized the rescue attempt was futile. The French, like the Marines, had headed for the mountainous area northeast of Beirut.

En route to Qartaba, the Marines' mechanized column encountered deep snow, blocked roads, and extremely difficult switchbacks. In many cases, amtracs had only part of their tread on the road. The rest

Photo courtesy of Claude Salhani

22d MAU assault amphibious vehicles laboriously make their way over narrow mountain roads on a rescue mission to recover Lebanese civilians stranded in the heavy snowstorm.

hung over the edge of the road, balanced presariously over a straight 300-400 foot drop into a deep gulley.[19]

On 24 February, the rescue mission ended. All units returned to their bases having won the gratitude of many Lebanese people. Later, after the snow began to melt, the Bekaa Valley flooded. Many Lebanese died and many others were placed in danger, and the Government of Lebanon borrowed some Marine equipment for its rescue efforts. The MAU loaned the LAF small boats from its reconnaissance platoon.[20]

These flood relief and rescue attempts by the three MNF units seemed to draw them more closely together. According to Colonel Mead, relations were cordial with the French, warm with the Italians, and neighborly with the British. In addition:

> There were more experiences facing us in the near term, however, that would bind the MNF commanders together in trust, respect, and friendship. Italian Gen [Franco] Angioni had been in Beirut since August; Gen [Michel] Datin, a French Marine, had relieved Gen [Jacques] Granger, a Legionnaire, in November; and British LtCol [John] Cochrane had only recently arrived. A most unique situation existed wherein there was no formal command relationship for the MNF. Our task became one of close coordination with each other, which was initially worrisome for old soldiers.[21]

Meanwhile, crosstraining continued with the Multi-National Force units as well as the LAF. The Italians participated in parachute jumps with members of the ANGLICO and Air Delivery Platoon detachments and the Phibron's underwater demolition team. Preparations were also made for future Marine Corps/French Marine crosstraining in amphibious operations.

Throughout February, Lieutenant Colonel Rice routinely attended a number of meetings at the American Embassy to exchange information and coordinate the MAU's activities with the embassy's. This was especially important in light of the strong diplomatic overtones of the Marines' mission. Rice also attended the twice-weekly MNF military coordination meetings at the Presidential Palace, and either he or Colonel Mead would attend the Political-Military Coordination Committee meeting, also held twice weekly at the Presidential Palace.

Colonel Mead hosted, briefed, and took VIPs on tours of MAU positions. On 17 February, Senator Roger W. Jepsen and his party arrived in Beirut. Two days later, Congressmen Jack Edwards, Anthony C. Beilenson, Carroll Campbell, Jr., Bernard J. Dwyer, Clarence

E. Miller, George M. O'Brien, Neal Smith, and Jack Hightower similarly landed at Beirut International Airport.[22]

The MAU's primary concern remained the terrorist threat. The primary need was for intelligence, more intelligence, and still more intelligence. Recognizing the danger that terrorists posed even before he led his command overseas, Colonel Mead requested from FMFLant a team of intelligence specialists to make an intelligence survey after the 22d MAU arrived in Beirut. The team arrived on 27 March and remained until 5 April.[23] In late April, when Vice Admiral Rowden, Sixth Fleet commander, visited his Task Force 61 and 62 commanders, a Sixth Fleet survey team accompanied him to review the intelligence setup.[24]

The Marines continued to perceive that their best defense was their posture of neutrality, so that a Muslim perception of U.S. neutrality *vis-a-vis* the Israelis and the Palestinians would be maintained. The Marines also recognized that the Shiites in Hay es Salaam seemed to prefer the situation that had existed since the American landings, and were being quite helpful. "They . . . helped out in providing some intelligence information which would help us in guarding against the terrorist threat."[25]

The shuttle diplomacy of Ambassadors Habib and Draper and Secretary of State George P. Shultz sought to obtain mutual agreement through which Syrian, Israeli, and PLO forces could leave Lebanon.[26] Marine Corps helicopters flew the diplomats from one Middle East capital to another. At the same time, Multi-National Force units began requiring from each other and from the LAF more intelligence reports from human sources (HUMINT). The MAU was becoming increasingly concerned about possible terrorist activities beyond the area immediately adjacent to its positions. The Marines recognized that in an urban terrorist environment, where conditions border on insurgency, HUMINT resources are invaluable. In the spring of 1983, however HUMINT information was just not available.[27] The MAU also maintained liaison with the Government of Lebanon, Lebanese intelligence services, Lebanese security forces, and Lebanese police, all of whom provided the Marines with bits of information. There was, nevertheless, a feeling that the Lebanese were not sharing everything they had.

The French MNF had its own human source intelligence network and provided the MAU with some intelligence that added substance to what the Marines already knew.[28] From the beginning, the MAU had multiple indications of being faced with an imminent terrorist threat. "An indication could be a telephone call, a letter . . . sent from one Lebanese organization to another, [or] a Lebanese civilian who would come forward with information."[29]

Based upon its mission and the limitations placed

The Multi-National Force commanders meet to discuss mutual problems of peacekeeping and defense, left to right: BGen Franco Angioni, Italy; Col James M. Mead, United States; BGen Michel Datin, France; and LtCol John C. Cochrane, Great Britain.

on the disposition of its forces by diplomatic considerations, the Marines could not go out to the countryside to confirm the threat reports. According to Major Farmer, considerable disinformation was published or broadcast by warring Lebanese factions. On numerous occasions, he was told that the MAU was going to be attacked by mortars or artillery at a given time. Such information was even broadcast over the local Phalangist radio station, the Voice of Lebanon.

The Marines consistently responded to these threats, knowing that they were vulnerable to terrorist attack. In order to conduct daily business, the MAU felt that it had to take this risk otherwise it might just as well have returned to its ships, or dug in deeply.

Lebanese sources usually provided the best HUMINT. It was impossible, however, to determine how much of this information was valid because the Marines had no feedback system for assessing the results of these actions. Marine response to HUMINT tips may have thwarted dozens of terrorists; or the Marines may have been batting near zero. They just couldn't tell which was the case.

Lieutenant Colonel Anderson, the BLT commander, was particularly concerned about the shortfall in HUMINT. He noted:

> My 2 [intelligence officer] can tell me what's going on in the Bekaa Valley and he can tell me what's going on in Tripoli, and he can tell me what's going on in this, that, and the other. We have no foggy idea of what's going on right outside our gate. We have no capability of tapping that and understanding how those people out there are feeling about us, if there's anything going on. That's one of my biggest problems and that is one of the things I don't know exactly how we solve.[30]

Marine response to threats consisted of reinforcing positions; restricting their activities outside of perimeter to patrolling only; increasing the number of patrols; digging in further; coordinating with the Lebanese for additional external security of MAU positions, for which the LAF was responsible; and coordinating with other MNF units for mutual security and mutual defense. The MAU would also increase its alert status and conditions.

In March, prompted by the stalled diplomatic negotiations, the 22d MAU augmented its airport perimeter defenses. The Marines watched the pace and results (or lack thereof) of the shuttle talks with some interest. No matter what transpired, they would be affected. Additionally, several other events occurred this month which had a bearing on the MAU's mission.

Despite the MAU awareness of increasing terrorist threats to its patrols, it continued sending them out daily. On 12 March, a foot patrol in Baabda was stopped at an Israeli checkpoint. The patrol leader was told that Marines were not supposed to be in that area. The Marines maneuvered around the checkpoint and continued their patrol. That afternoon, a second patrol in Baabda was challenged by the Israelis just before it re-entered Marine positions. As that patrol, too, maneuvered around the checkpoint and returned to friendly lines, it was verbally harassed by the Israelis.[31] These events were reported up the chain of command. It was obvious that these confrontations could not continue and Colonel Mead took the matter up with Ambassador Habib.

Mr. Habib then met with Israeli Minister of Defense Moshe Arens, and told him that the MAU commander would personally lead the patrol through the Israeli checkpoint the next time it tried to stop a Marine in the execution of his duties. The Baabda patrols were temporarily suspended for several days until the matter of challenges were clarifed.[32]

Though not in the operational chain of command leading down to the MAU (except as a member of the Joint Chiefs of Staff), General Barrow, Commandant of the Marine Corps, closely monitored all developments in Lebanon. The latest harassment of the Marines compelled him to write a letter on 14 March to the Secretary of Defense in which he demanded that ". . . firm and strong action" be taken to stop Israeli forces from putting the Marine and Army officers in ". . . life threatening situations that are timed, orchestrated, and executed for obtuse Israeli political purposes." General Barrow was concerned not only with the harassment of the Marine patrols but also the threats to Marine and Army officers assigned to the United Nations Truce Supervisory Organization in Lebanon.

The Secretary of Defense supported General Barrow's position by forwarding the Commandant's complaint to Secretary of State Shultz, who took the matter up with Israeli Foreign Minister Yitzhak Shamir, then visiting in Washington.

On 24 March, following publication of the Commandant's letter and diplomatic representations made to the Israeli government, Colonel Mead, Deputy Chief of Mission Pugh, and Marine Colonel Cornwill R. Casey, the EUCOM liaison officer, met with Israeli General Lifkin, and ". . . discussed the exchange of patrol information between 22d MAU and the IDF as a possible solution to avoiding further contacts."[33] Once this procedure was adopted, there were few further incidents.

A telephone line was installed linking the MAU po-

Photo courtesy of Claude Salhani

22d MAU Marines of a BLT 2/6 patrol, rifles at the ready and rounds chambered, take cover behind construction equipment on the coastal road south of Beirut after having been attacked on 16 March 1983. The white-helmeted soldier and the one with dark glasses are soldiers belonging to the French contingent, United Nations Interim Force in Lebanon (UNIFIL). They happened to be passing by at the time of the incident.

sition at Lebanese University with the Israeli company position across the Sidon Road. A direct radio/telephone link was established between Colonel Mead and General Lifkin.[34] This was in addition to the IDF emergency radio net, already manned by all MNF contingents.

In reporting to FMFLant, Colonel Mead noted that this meeting with General Lifkin had been serious and professional and that the Marine Corps-Israeli problems in Beirut were apparently defused. More ominously he noted that "[Terrorist] threat increases as diplomatic situation stagnates."[35]

Whether or not a cause-and-effect relationship existed with the diplomatic situation, the terrorist threat seemed to escalate in March, influenced by several factors. The Syrians, in the wake of their earlier battering by the Israelis, had been re-equipped by the Soviet Union with better and more modern weapons. Syrian troops had been re-trained and reinforced. In addition, PLO gunmen had infiltrated into the Beirut area. Weather conditions had also improved by the end of March, providing a more congenial climate for terrorist activities. Marine staff officers anticipated a change in the threat for it appeared that individual units, organizations, and sects in the Beirut area were girding up to protect their vested interests. The terrorist acts that were to come in succeeding months were ". . . symptoms of what was actually taking place organizationally to the infrastructure of the local Palestinian sects and organizations."[36]

Meanwhile, Lebanese reaction to the presence of the MNF in Beirut ranged from total acceptance to relative indifference, or so it seemed. A harbinger of change came dramatically on the night of 15 March, however, when an Italian mobile patrol was ambushed by persons unknown. One Italian soldier was killed and nine others wounded. The next day, a hand grenade was thrown from a second-story window of an apartment at a Marine foot patrol in Ouzai, north of the north-south runway of Beirut International Airport. Five Marines sustained superficial wounds, and were helicoptered to the *Guadalcanal* for treatment, subsequently returning to duty. On 18 March, General Tannous decorated each with the Lebanese Medal of War and the Medal of Injury.

Shortly after receiving word of the attack, a Marine reaction force was on the scene, as were Italian and LAF troops. It was not immediately determined which faction had attacked the Marines, but the Lebanese arrested more than 100 individuals. Subsequently, a Lebanese citizen who supported the Amal faction was tried and convicted of the grenade attack, and sentenced to death.[37]

The Marine force was next activated in the early morning hours of 17 March at the request of the Italian MNF to seal off the airport after the Italian command post was hit by small arms fire. This was still another example of the closer relationship being engendered between Multi-National Force contingents. In April, there would be a greater demonstration of this international camaraderie.

The Marines—and the French, Italians, and British—were now facing an increased threat to the Multi-National Force. On 25 March, 22d MAU Marines began conducting all patrols with loaded magazines inserted in their weapons, as authorized by the Rules of Engagement issued in February. To tighten their security, as well, all mobile patrol routes were alerted on 27 March. Three days later, the MAU suspended foot patrols in Baabda, but at the same time, it increased mobile patrols in that area.[38]

In March, Lieutenant General John H. Miller, commanding general of FMFLant, the MAU's parent command in the Marine Corps chain, arrived in Beirut. With his party was Major General Bernard E. Trainor, the Director of the Plans Division and soon to be Assistant Chief of Staff for Plans, Policies, and Operations at Headquarters Marine Corps. The purpose of the visit was to talk with Colonel Mead, Ambassador Dillon, and General Tannous, and visit the Marines of the MNF. Later in the month, Colonel Mead was visited by additional congressional parties, including Representatives Lyle Williams on the 25th, and Geraldine A. Ferraro and Barbara A. Mikulski on the 29th, and Patricia Schroeder on the 31st.

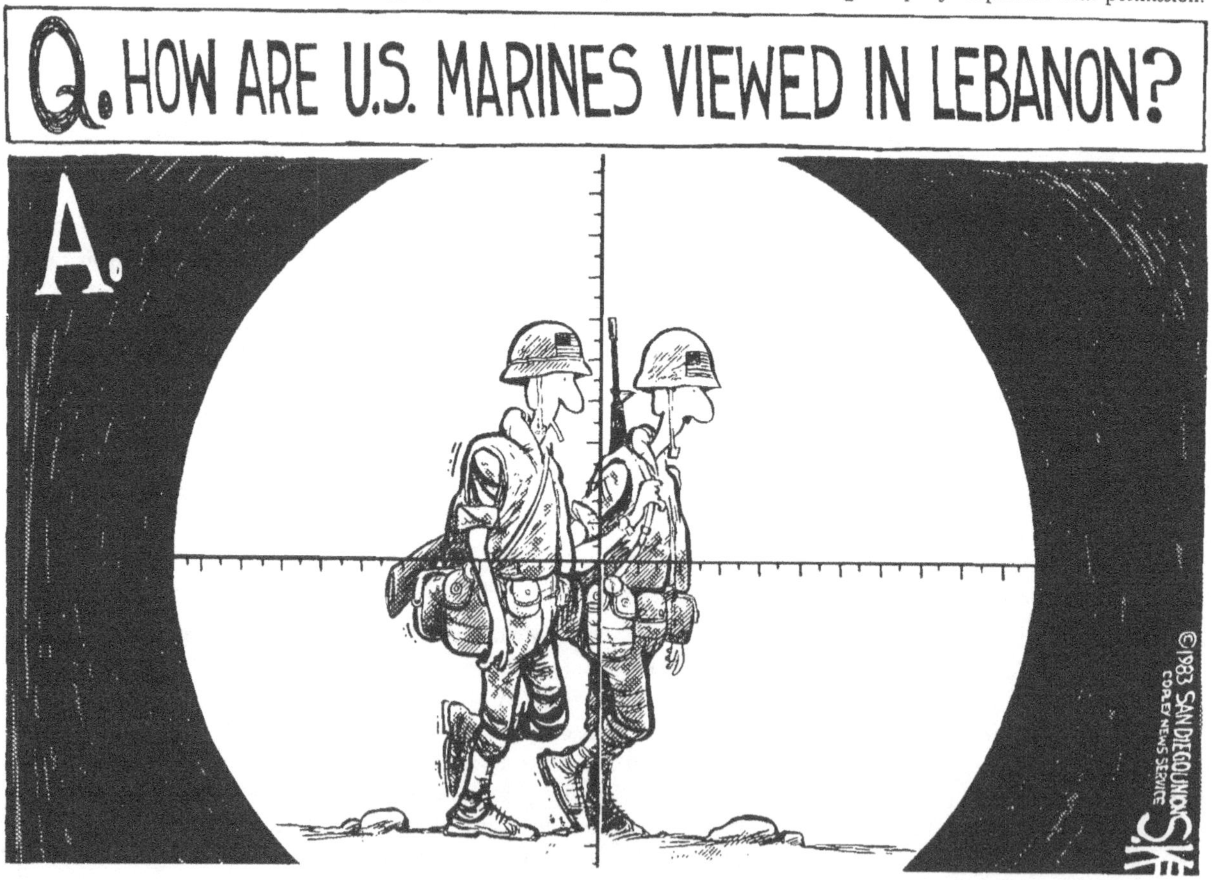

Photo courtesy of Claude Salhani

Hospital Corpsman 2d Class Donald E. Sagamang is hugged by Ali Mohamed in the Hay es Salaam area of Beirut. The Navy medic saved the elderly Lebanese man's life after he suffered a heart attack when the corpsman was on patrol with a unit from BLT 2/6. On 29 March 1983, during a subsequent patrol, Mohamed recognized and embraced the American sailor.

Throughout the month, the Marines continued training LAF units and cross-training with the French Marines, while HMM-264 pilots flew "Cammie Cab Company" missions between Israel and Lebanon, and between Beirut and Larnaca. The 22d MAU also continued the medical community assistance program established earlier by the 24th MAU. By the end of the month, Navy medical personnel had treated 279 Lebanese patients.

In his weekly report to FMFLant for the period 2-8 April, Colonel Mead commented that the situation in Beirut had become unusually quiet.[39] In the first weeks of the month, he hosted and provided orientation briefings to Congressman Louis Stokes and the Deputy CinCEur, General William Y. Smith. The situation didn't remain quiet for long, however. April also brought terrorist attacks on the French and Italian contingents of the MNF, resulting in the death of another Italian soldier. On the night of the 17th, a Marine sentry on duty at the Company F command post was fired upon by an unknown assailant. For the first time in the deployments, Marines returned fire, but with unknown results. The enemy round ripped off the cargo pocket of the Marine's utility trousers, but luckily missed his leg.

The sense of accomplishment following the appar-

The American Embassy in west Beirut, as it appeared before the April 1983 bombing.

Photo courtesy Marine Security Guard School

Photo courtesy of Claude Salhani

An aerial view of the American Embassy as heavy cranes continue to remove rubble from the upper floors on 21 April, following the terrorist bombing three days earlier.

Nighttime clearing operations at the Embassy. The arrow points to Ambassador Dillon's top-floor office, which he occupied when the bomb exploded.

USMC Photo by GySgt Jimmy C. Hickman

ent successes of Beiruts I and II was diminishing in the face of the increasing terrorist threat. But no one expected the catastrophic bombing of the American Embassy in Beirut on 18 April. This vicious act was perpetrated by a terrorist driving a van, reportedly stolen from the Embassy in June 1982. The van carried a 2,000-pound load of explosives. Once the driver had driven past a lone sleeping Lebanese guard, he came to a halt in the lobby of the building where the van exploded with great force. The blast tore through the front portion of the seven-story Embassy, killing 63 occupants. These included 17 Americans, one of whom was Corporal Robert V. McMaugh, a member of the Marine Security Guard detachment at the embassy, who was manning Post No. 1 in the main lobby.

While it undoubtedly heard the explosion, being only about four miles from the Embassy, the MAU headquarters staff first learned of the bombing when a staff sergeant from the S-3 Section, who had just left the Embassy two minutes before the explosion, called to report what had happened. Shortly thereafter, BLT 2/6 provided a reaction company to secure the area. All available corpsmen were also dispatched to the Embassy to treat the injured, while two surgi-

USMC Photo by GySgt Jimmy C. Hickman

A somewhat disheveled Sgt Luis G. López, of the Marine Security Guard detachment, stands guard outside of the Embassy immediately after its bombing.

cal teams from the Phibron were sent ashore to assist the staff of American University Hospital.

Colonel Mead arrived within 20 minutes after the explosion to find that General Datin, in whose sector the Embassy was located, was already on the scene guarding it with his French Marines. they had cordoned off the area and had begun initial rescue efforts. General Datin generously offered to place his men under the operational control of Colonel Mead, an offer graciously made and gratefully accepted. As soon as the U.S. Marine reaction company arrived, it relieved the French and took up the security detail. For the next two weeks, the MAU Marines were to be kept quite busy.

The MAU and then the LAF joined the French in the area cordon. The BLT established a 24-hour security guard as the rescue and clean-up operations continued. All MAU Marines on shore served in security or logistical duties at some time or another outside the Embassy because they were needed and also to bring the realities of terrorism to them.

Some of the MAU Marines did not get more than three or four hours of sleep a night during the two-week period immediately following the blast. They would get up to go on patrol, return to their positions, ". . . and might get one or two hours to take care of personal matters and so on, have chow, and go back out on the lines, standing post around the security effort at the embassy site."[40]

The initial reaction of the Marines to this disaster when they arrived at the site and saw the devastation was one of absolute anger. Said Colonel Mead later, they also asked, "'How dare anyone strike the U.S. Embassy?' They were angry. And . . . they were appalled by the magnitude. They were frustrated because they couldn't do anything about it. And who do you lash out at? Then the next reaction was, 'Gee, this could happen to me.'"[41] And then, according to Colonel Mead, the Marines ". . . dealt with their own mortali-

American Ambassador to Lebanon Robert S. Dillon points out bomb damage to Under Secretary of State Lawrence S. Eagleburger as Col Mead looks on.

Department of Defense (USMC) Photo

Copyright 1983. Pat Oliphant. *Universal Press Syndicate.* Reprinted with permission.

ty, and they matured almost on the spot, each Marine."[42]

There wasn't much time for such introspection, however, as one company (-) was pressed into securing a perimeter immediately around the remains of the Embassy building. Together with the Embassy's Marine security guard, 22d MAU Marines methodically searched the rubble to recover whatever classified material could be found and to verify that none was left after the search. Most of the security guard detachment had been in the building when the bomb exploded and were temporarily stunned. Nonetheless, they "performed magnificently" according to General Mead.[43] The next morning, at 0500, members of the detachment raised the American flag over the Embassy's ruins at Ambassador Dillon's request.

The MAU was somewhat shorthanded at this time, because 136 Marines were on the *Raleigh* for a port visit to Athens, and 125 more were at Camp Des Garrigues, a French Army base near Numes, France for training and liberty. The two groups rejoined the MAU on the 16th and 18th respectively.

Soon after the dust of the explosion settled, and it became apparent that the building was no longer habitable, British Ambassador Sir David Roberts, acting on his own authority, offered working spaces in the British Embassy for the American Embassy's political, military, and consular sections. The British Embassy was located on the Corniche, a major Beirut

Marine guard post outside of the British Embassy.

USMC Photo by GySgt Jimmy C. Hickman

Nighttime security post established on the Corniche outside the Durrafourd Building in April 1983.

thoroughfare along the Mediterranean. Just down the road, several buildings away was the Durrafourd Building, where the rest of the American diplomats would set up temporary headquarters.

Sir David also requested that the MAU provide security for the British Embassy, marking ". . . probably the first time in history that you have U.S. Marines guarding a British Embassy."[44] Colonel Mead responded by ordering a platoon to the site. The Marines were commanded by Lieutenant William G. Leftwich III, whose father, a Marine lieutenant colonel, had been posthumously awarded the Navy Cross during the Vietnam War.[45] The platoon also set up sandbagged guard posts outside the Durrafourd Building. One Marine platoon joined embassy Marines in providing security for Ambassador Dillon's home in Yarze, in the hills just east of Beirut. The MAU also provided overnight security to a joint State Department/Central Intelligence Agency delegation, which arrived in Beirut on 22 April to escort the bodies of the dead Americans home.

With the establishment of the security forces at the British Embassy, and for a while at the devastated American Embassy, the MAU's mission changed. In addition to its earlier mission of presence, it now had one of providing security. The MAU changed the rules of engagement to permit a Marine to fire if he "perceived" hostile intent. The new rules were, like the

Marine guard post outside MAU headquarters at Beirut Airport at the end of May 1983.

Photograph by the author

ones then in force, printed on a wallet-sized blue card and issued to the Marine standing guard at the embassies.

These new rules were as follows:

Rules of Engagement for American and British Embassy External Security Forces

1. Loaded magazines will be in weapons at all times when on post, bolt closed, weapon on safe. No round will be in the chamber.

2. Round will be chambered only when intending to fire.

3. Weapon will be fired only under the following circumstances:

a. A hostile act has been committed.

(1) A hostile act is defined as rounds fired at the embassy, embassy personnel, embassy vehicle, or Marine sentries.

(2) The response will be proportional.

(3) The response will cease when attack ceases.

(4) There will be no pursuit by fire.

(5) A hostile act from a vehicle is when it crosses the established barricade. First fire to disable the vehicle and apprehend occupants. If the vehicle cannot be stopped, fire at the occupants.

(6) A hostile act from an individual or group of individuals is present when they cross the barricade and will not stop after warnings in Arabic and French. If they do not stop, fire at them.

4. Well aimed fire will be used; weapons will not be placed on automatic.

5. Care will be taken to avoid civilian casualties.

U.S. Secretary of State George P. Shultz (c), is greeted by 22d MAU commander Col James M. Mead (l) and Ambassador Robert S. Dillon (r) on 28 April 1983, when Shultz arrived at Beirut International Airport for a meeting with Lebanese President Gemayel that day.

Photo courtesy of Claude Salhani

Copyright 1983. Paul Conrad. *Los Angles Times.* Reprinted with permission.

In his report to General Miller, Colonel Mead commented on his new situation by saying, "In spite of the terrorist threat, we are continuing to maintain a proper balance between our security and our presence/peacekeeping mission."[46]

The new rules of engagement were tested at 0200 on 28 April, when two men in an automobile tried to run the joint LAF/Marine checkpoint at the British Embassy. Three warning shots were fired, and then three more shots, forcing the vehicle off the road. The Lebanese apprehended the two men, who appeared intoxicated. One turned out to be a Syrian national with faked Lebanese identification papers in his possession. The word quickly went out that the Marines will shoot back and that they hit what they shoot at, ". . . and that gave . . . a little bit more credibility to our [fighting] ability and what we were willing to do in regards to the security situation here for the Lebanese; and it also gave a warning to the terrorists."[47]

As the Beirut environment began to change dramatically for the Marines, the MAU took further steps to strengthen its positions. Additional barriers were constructed in the airport area; sentries were doubled at all posts; and all vehicles were subjected to even more detailed searches before they were permitted to enter the MAU compound. Colonel Mead requested an on-the-scene intelligence team to coordinate all

Photograph by the author

A stray .50 caliber round hit this tree outside the Joint Public Affairs Bureau office in the 22d MAU compound. The Marine combat correspondents assigned to the MAU awarded it a Purple Heart of sorts.

American intelligence efforts, to sort through the great amount of intelligence material available, and to assess the threats as soon as they were perceived.[48]

Tanks, AAVs, and artillery, which had been brought ashore in February for maintenance and training, were now employed in other ways. The amtracs were used to block routes coming within the MAU areas, tanks were placed in a centralized location in the MAU perimeter for rapid response missions and the artillery battery began a more active target acquisition effort.

With diplomatic negotiations lagging, Secretary of State Shultz arrived in Beirut on 28 April to lend his efforts to the peacekeeping negotiations. The MAU provided security and transportation for Secretary Shultz as well as JCS Chairman General John W. Vessey and Sixth Fleet commander Vice Admiral Rowden, who arrived on the same day. Before his return to the States, Mr. Shultz sent the following message to the Commandant on 4 May:

Dear General Barrow:

Over the last week, I have made three visits to Beirut. On each occasion the 22d MAU under Colonel Mead provided unstinting support in security and transportation arrangements. I have also had the opportunity to observe these Marines in the difficult and dangerous circumstances of Beirut. They made a fine, sharp outfit. They are disciplined, professional and spirited. The Marines are highly regarded by their

> Lebanese hosts and by their colleagues in the MNF. I have always been proud to be a Marine, but never prouder.
>
> George Shultz[49]

The beginning of May brought another obvious and dramatic shift in conditions around Beirut. Between 5 and 8 May, fighting among the Christian Lebanese Forces militia, the Phalangists, and the Muslim Druze spilled over into Beirut in the form of artillery shelling. During the worst of the bombardment, rockets and artillery hit Juniyah, Muslim west Beirut, and Christian east Beirut. On 5 May, when it appeared that the French position was being attacked, the MAU put its artillery, mortars, Cobra gunships, and naval gunfire assets on alert. Since the MNF rules of engagement stipulated that an attack on one constituted an attack on all, Colonel Mead went aloft in a HMM-264 Huey with two aerial observers to locate the battery that was firing on the French. At the same time, the MAU S-3 sent ANGLICO artillery and naval gunfire spotters to observation posts in the Baabda hills above the airport. Colonel Mead's aircraft was hit by three 7.62mm rounds, which did no severe damage. The artillery firing on the French sector ceased at about 1327 and the MAU stood down from its alert posture at 1430.[50]

At 2025 on 6 May, intelligence sources reported that Druze artillery would shell the airport and MAU positions that night. All MAU artillery and mortars were alerted, as were the naval gunfire support ships. HMM-264 aircraft still at the airfield returned to the *Guadalcanal*. During the evening, artillery and rocket rounds impacted in the French and Italian sectors with no MNF casualties (there were civilian casualties), but the airport was not hit. All supporting arms relaxed their alert status at 0240 on 7 May. Intermittent small arms fire continued through the early morning hours of the 7th in the hills to the east of the airport, with stray rounds impacting in some of the Marine positions, but no Marines were hurt.[51]

At 1447 that afternoon, a round landed inside the Marine positions at the southern end of the runway. Minutes later, several more rounds landed in the sea, 500-1,000 meters from the beach. It was soon determined that the firing came from two locations immediately to the east of the Marine lines. At 1455, the artillery and mortars were put back on alert and a rifle company was ordered to prepare for a sweep of the area southeast of the airport between the railroad tracks and Old Sidon Road. Meanwhile, a Huey command-and-control helicopter was launched to make a reconnaissance of the area. The Israelis were alerted (via the emergency net) that the Marines intended to make a ground sweep, unless the Israelis investigated the suspected firing positions first. The

Col Mead briefs Commandant of the Marine Corps Gen Robert H. Barrow during the latter's farewell visit to 22d MAU in Lebanon on 26 May 1983. Note sandbags outside of the windows of the 22d MAU commander's second-story office at Beirut Airport.

Photograph by the author

Photograph by the author

The old and the new: (l. to r.) Capt Morgan M. France, USN, ComPhibron 8, relieved Capt George Bess, USN, ComPhibron 2, on 29 May 1983, while Col Mead, commander of 22d MAU was relieved the same day by 24th MAU commander, Col Geraghty.

Sgt Charles A. Light, NCOIC of the Beirut Embassy Marine Security Guard detachment, presents a plaque of appreciation from the detachment to Col Mead on 29 May 1983.

Photograph by the author

Col Mead honors Gen Tannous at the 22d Marine Amphibious Unit mess night hosting Multi-National Force officers. At the left is BGen Datin and on the right, BGen Angioni.

IDF responded that they would investigate the area, and did so without results.

The shelling and rocketing continued, and inasmuch as Secretary Shultz was scheduled to arrive at the airport on the 8th, the MAU was concerned. The majority of the shelling was now landing close by—directly across from the airport in Kfar Shima, a Christian sector, and Ash Shuwayfat, held by the Muslim Druze. The shelling continued the next day and then diminished and finally ceased about 40 minutes before the Secretary landed in Beirut.[52]

The beginning of the end of the 22d MAU's stay in Beirut neared on 10 May with a re-embarkation conference held with Commodore Bess and his staff. Four days later, the 24th MAU's executive officer, Lieutenant Colonel William A. Beebe II, together with the 24th's S-4, Major Robert S. Melton, arrived in Beirut to begin coordinating the relief in place. On 19 May, the 22d MAU security element at the devastated embassy site returned to its parent unit.

Two days earlier, on 17 May, Israel and the Government of Lebanon had signed an agreement calling for the withdrawal of Israeli forces from the Chouf region and for the institution of special measures in southern Lebanon to guarantee Israel's security. As the Long Commission noted:

> Israel, however, predicated is own withdrawal on the simultaneous withdrawal of Syrian and Palestinian Liberation Organization (PLO) forces from Lebanon, parties which had not been included in the negotiations. Syria refused to initiate withdrawal of its forces while the IDF remained in Lebanon. The stage was set for renewed violence.[53]

This was the situation the 24th MAU would enter when it came ashore on 29 May. Before the 22d left, it was visited on 26 May by the Commandant, General Robert H. Barrow, who was making farewell visits to Marines around the world prior to his retirement on 1 July after more than 40 years of active service and participation in three wars. General Barrow visited all the MAU positions and presented Purple Hearts to the five BLT 2/6 Marines who had been wounded in the grenade attack early in the deployment. He also visited the American Embassy site and then the British Embassy. The Commandant met with General Tannous at the Ministry of Defense, and then he visited the French contingent, where, on behalf of the President, he decorated General Datin with the Legion of Merit for his assistance in the aftermath of the embassy bombing. Nine other French officers and medical personnel were also decorated for their services.

Despite his short period in Lebanon, the Commandant spent some considerable time in conference with Colonel Mead; Colonel Timothy J. Geraghty, the incoming 24th MAU commander; Colonel James P. "Pat" Faulkner, who was to relieve Colonel Mead in June at Camp Lejeune as commander of the 22d MAU; Commodore Bess; and Colonel Jim R. Joy, then Sixth Fleet Marine Officer. General Barrow commented on his perceptions of the MAU situation and positions. He strongly emphasized one point, among several, and that was American, and Marine, HUMINT was poor and that it had to be improved—now.[54]

On 27 May, at ceremonies held at the Ministry of Defense, General Tannous decorated Commodore Bess, Colonel Mead, Lieutenant Colonels Rice, Anderson, and Kalata, and Major Shively with the Lebanese Order of the Cedar Medal. On that same day, General Barrow returned to the United States and Congressmen William H. Gray and Stephen J. Solarz visited the Marines.

The advance party of the 24th MAU had arrived in Beirut on 24 May. Four days later, the new MAU's advance command, control, and communications elements arrived to reconnoiter the area. The 24th MAU Marines also assumed the security detail at the British Embassy and the Durrafourd Building. On 29 May, the major elements of the 24th MAU began landing at 0700. They were in their new positions by 1300, at which time the 22d MAU's units began reembarking. At 1600 on the 30th, Colonel Timothy J. Geraghty, the 24th MAU commander assumed responsibility as commander of the U.S. contingent of the Multi-National Force in Beirut. Phibron 2 and the 22d MAU were on their way to Rota to wash down their equipment and enjoy two or three days of liberty. Following that, they were headed for Key West and Operation Agile Retrieval on 10 June.

After three and a half months ashore in Lebanon, the 22d MAU left with a sense of solid accomplishment. The BLT's Marines ". . . dealt with patrols, security, terrorist attacks, rescue operations, and cross training requirements—all with the ease of practiced professionalism."[55] Meanwhile, Lieutenant Colonel Kalata's HMM-264 flew 2,303 hours with an aircraft availability of 94 percent. In this deployment, the choppers transported 19,200 passengers, hauled 988.4 tons of cargo, and fully supported the U.S. diplomatic mission. Major Shively's MAU service support group had come ashore with 300 pieces of rolling stock and managed to maintain a 96.6 percent operational rate, while meeting all MAU logistics requirements.[56]

On the Thursday evening before the 22d MAU left Beirut, Colonel Mead hosted a mess night for the officers of the MAU and the MNF units in Lebanon. At the end of a candlelit "surf and turf" dinner of steak, lobster, and wines, when the toasts were made and the VIPs introduced, Colonel Mead made a presentation to General Tannous on behalf of the MAU. As General Tannous rose to make his remarks, a heavy caliber round exploded just outside of the MAU perimeter. General Tannous smiled, there was some laughter, and one who was there was reminded of an old French aphorism, "The more things change, the more they remain the same."[57]

On 27 June, at Camp Lejeune Colonel Mead turned over command of the 22d Marine Amphibious Unit to another veteran aviator, Colonel James P. Faulkner, whose 22d MAU was destined to make Marine Corps history before it ever reached Beirut.

CHAPTER 6

Beirut V—Disaster Strikes 30 May-19 November 1983

Colonel Timothy J. Geraghty, who had assumed command of the 24th MAU on 17 March 1983, reported for operational control to the CG, Fleet Marine Force, Atlantic, 24 March.[1] On that same day, BLT 1/8 (Lieutenant Colonel Howard L. Gerlach), HMM-162 (Lieutenant Colonel Laurence R. Medlin), and MSSG-24 (Major Douglas C. Redlich) reported for operations to the 24th MAU. For the upcoming deployment, the 24th MAU would go to Lebanon with Commodore (Captain, USN) Morgan M. France's Amphibious Squadron 8 on board the *Iwo Jima* (LPH 2), the Phibron flagship; the *Austin* (LPD 4); the *Portland* (LSD 37); the *Harlan County* (LST 1196); and the combat cargo ship *El Paso* (LKA 117).

MAU planning and operations were governed by an FMFLant letter of instruction, which had been issued in early February.[2] The letter provided the MAU with instructions concerning its mission; operational, administrative, and logistics matters; and command and communications matters. It resembled the Marines' standard five-paragraph operations order, and was similar to letters of instruction issued to earlier MAUs.

On 27 April, the 24th's advance liaison party left for Lebanon. It returned to Camp Lejeune on 2 May to give orientation briefings to the commander and staff. The MAU embarked on Phibron shipping at Morehead City and Onslow Beach on 11 May and headed east toward the Mediterranean, but MSSG 24 did not leave Morehead City until the next day. The *El Paso*, upon which the MSSG was embarked, had engine trouble that was not repaired until the 12th.

Two days earlier, Colonel Geraghty had visited Washington for briefings at Headquarters, U.S. Marine Corps, and the Department of State. His staff went to Fort Bragg, North Carolina, for a Lebanon briefing by the Army's 8th Psychological Operations Battalion. On the trip across the Atlantic, the MAU commander and his staff visited all the ships in the Phibron, and Colonel Geraghty also gave a three-hour personal briefing to embarked Marines and Phibron crew members in which he covered the Marine Air-Ground Task Force organization; 24th MAU organization; the history, religions, politics, and social culture of the Lebanese; the foreign and domestic factions in Lebanon; the rules of land warfare and of engagement; public affairs matters; and naval intelligence and operations.

During the Atlantic transit, a young Marine suffered an acute attack of appendicitis. Because there was no anesthesiologist in the MARG, in mid-Atlantic he had to be helilifted for surgery to the carrier *Nimitz*, then heading west to the States. Colonel Geraghty noted that this highlighted a serious medical deficiency in the MARG, and recommended that the ". . . possibility of obtaining necessary qualified medical personnel be explored to insure availability to all MARGs transiting the Atlantic."[3]

The 24th MAU differed from previously deployed MAUs in several ways. For example, the personnel assigned to Colonel Geraghty's staff were the first to be assigned permanently on change of station orders to

Before leaving for Beirut, BLT 1/8 commander LtCol Howard L. Gerlach, inspects an 81mm mortar with an M-32 sub-caliber pneumatic trainer attached.

USMC Photo by SSgt Robert E. Kline

An HMM-162 CH-53E "Super Stallion" lifts off from the deck of the Iwo Jima, *laying off the beach near Beirut International Airport, and heads inland to support Marines ashore.*

a Marine Amphibious Unit headquarters for a tour of duty. Previously, when a deployment ended and a MAU returned to Camp Lejeune, its TAD personnel returned to their parent commands on the base. This welcome change offered ". . . continuity and a corporate memory . . . particularly for the sustained operations we have now in Lebanon."[4]

Another first in this deployment was the fact that HMM-162 was equipped with the new CH-53E "Super Stallion," a more versatile helicopter than the previous model, the CH-53D, and with a capability of lifting 16 tons. This was an especially important factor, for the 24th MAU was also deploying with the new M-198 155mm towed howitzer, which could only be helilifted by the CH-53E. During the predeployment period, HMM-162's heavy-lift helicopters extensively trained lifting the howitzers from the deck of the *Austin*.

Other new equipment introduced for specific use ashore in Beirut included the Reverse Osmosis Water Purification Unit (ROWPU), for making fresh water, and the Mobile Food Service Unit, which is used when tray pack foods, a new concept in field rations, were issued to the troops. The 24th noted, "both units work extremely well and have proved to be major morale factors for the Marines ashore."[5]

After entering the Mediterranean, the Navy and Marine officers were given an updated Beirut brief by Sixth Fleet Marine Officer Colonel Jim R. Joy on 25 May. The next day, Colonel Geraghty and his advance party flew ashore to Souda Bay, Crete, and from there on to Beirut.

The relief in place of the 22d MAU was carried out with no perceptible problems on 29 May. Once in position, the 24th MAU immediately began mobile and foot patrols and took up positions at the airport and at the security post guarding the Durrafourd Building and the American/British Embassy. Colonel Geraghty was in Beirut less than a week before the first VIPs visited him. On 2 June, he and Commodore France hosted and briefed Congressmen Thomas M. Foglietta, Peter H. Kostmayer, and Theodore S. Weiss. Six days later, Vice Admiral Rowden arrived to meet the new commanders of Task Forces 61 and 62. Towards the end of the month, Chapman B. Cox, Assistant Secretary of the Navy for Manpower and Reserve Affairs, and a reserve Marine lieutenant colonel, arrived on the 28th for a two-day visit and orientation.

Like the MAUs before it, the 24th MAU set up its headquarters in the airport fire fighter school. The BLT headquarters and attached units established themselves in the four-story building that once had housed the Government of Lebanon's Aviation Administration Bureau. In picking its command post, the BLT sought a site that provided security from light to heavy hostile artillery, rocket, and sniper fire, which had caused few Marines casualties so far.

Upon assuming airport defensive positions previously manned by the 22d MAU, BLT 1/8 began immediately to improve them. From 29 May—when Lieutenant Colonel Gerlach's battalion landed—to 23 October of that year, his Marines filled some 500,000 sandbags and emplaced 1,000 engineer stakes and 10,000 feet of concertina wire.

As the Long Commission later determined:

> The BLT Headquarters building was occupied from the outset for a variety of reasons. The steel and reinforced concrete construction of the BLT Headquarters building was viewed as providing ideal protection from a variety of weapons. The building also afforded several military advantages that could be gained nowhere else within the BLT's assigned area of responsibility. First, it provided an ideal location to effectively support a BLT on a day-to-day basis. Logistic support was centrally located, thus enabling water, rations, and ammunition to be easily allocated from a single, central point to the rifle companies and attached units. The Battalion Aid Station could be safeguarded in a clean, habitable location that could be quickly and easily reached. Motor transport assets could be parked and maintained in a common motor pool area. A reaction force could be mustered in a protected area and held in readiness for emergencies. The building also provided a safe and convenient location to brief the large number of U.S. Congressmen, Administration officials, and the flag and general officers who visited Beirut from September 1982 to October 1983. In sum, the building was an ideal location for the command post of a battalion actively engaged in fulfilling a peace-keeping and presence mission.
>
> Second, the building was an excellent observation post. From its rooftop, a full 360 degree field of vision was available. From this elevated position, forward air controllers, naval gunfire spotters and artillery forward observers could see into the critical Shuf [Chouf] Mountains area. Also from this position, observers could see and assist USMNF units in their positions at the Lebanese Science and Technical University. Further, this observation position facilitated control of helicopter landing zones that were critical to resupply and medical evacuation for the MAU. In sum, many of the key command and control functions essential to the well-being of the USMNF as a whole could be carried out from the building. No other site was available within the bounds of the airport area which afforded these advantages.
>
> Third, the building provided an excellent platform upon which communications antennae could be mounted. In that the supporting ships were initially as far as 3,000 to 6,000 yards off shore, antenna height was a major factor in maintaining reliable communications with the supporting elements of the 6th Fleet. Reliable communication with [CTF 61 and the ships of CTF 60] was critical to the defense and safety of not only the USMNF, but to the U.S. Embassy, the U.S. Ambassador's residence, the Durrafourd building, and our allies in the MNF as well. Reliable communications meant that naval gunfire missions could be directed at hostile artillery and rocket positions in the Shuf Mountains when they fired into the airport. Line-of-sight communications are also essential in calling for and adjusting air strikes. Moreover, such communications were key to the rapid evacuation of casualties via helicopter to secure medical facilities offshore.[6]

The relative quiet of late May and most of the following month permitted the 24th MAU to send a group of Marines to France for training and liberty and other Marines to Athens and Turkey for liberty, without weakening the MAU's readiness posture. A composite company of 102 Marines and sailors from

Marines from the 24th MAU dangle from a rope attached to a CH-46 helicopter over Beirut in a joint SPIE (special patrol insert-extract) rig exercise on 24 August 1983 with Legionnaires serving with the French contingent of the Multi-National Force.

Photo courtesy of Claude Salhani

the BLT went to Camp Des Garrigues, 102 more went to Athens on board the *Harlan County*, and at the end of the month, 192 went to Antalya, Turkey on board the *Portland* for liberty.[7]

The MAU continued sending out an average of four to seven foot patrols daily in the vicinity of the airport, varying the sites, times, and routes. As the area outside the airport was becoming increasingly hostile, Colonel Geraghty met with the MNF Military Committee on 20 June to propose that Lebanese Army Forces fire teams accompany the Marines in the hope that adding LAF soldiers to the patrols might alleviate the threat.[8] Five days later, such a step was taken. Each mobile patrol was also accompanied by a LAF lieutenant.

The Israelis continued to have their problems with terrorists. On 23 June, they responded to hostile acts with small arms fire that landed within Marine positions. The MAU commander took the matter up with the EUCOM representative and Israeli officers.

In June, the MAU conducted a heavy schedule of crosstraining, including SPIE [Special patrol insert/extract] rig/rappelling training with the LAF air assault battalion and soldiers from The Queen's 1st Dragoon Guards, the British MNF contingent. On 26 June, Marines and members of the French unit, fired each other's weapons to become familiar with them. Throughout the month, the MAU's mechanics provided diesel engine training to their LAF counterparts. At the same time, all subordinate commands within the 24th MAU conducted a comprehensive training schedule that concerned physical fitness, small unit tactics, leadership, troop information, and field sanitation. The helicopter squadron also had a heavy training schedule, which included an extensive program of day and night qualification flights. Since its arrival in Beirut, Lieutenant Colonel Medlin had provided two aircraft on continuous medevac alert, as well as two Cobras on strip alert. While flying a round of logistics and diplomatic flights, HMM-162 also participated in the MAU's crosstraining program by providing support for air assault operations, reconnaissance inserts, and helicopter familiarization.

On the intelligence side, the MAU's S-2 section held

A Cobra from HMM-162 rises from the Iwo Jima *to support the 24th Marine Amphibious Unit ashore. For their security, Cobras were generally kept on board their carriers.*

USMC Photo by SSgt Robert E. Kline

security and antiterrorist countermeasures classes for all MAU commands. The section also developed an extensive and comprehensive intelligence and security awareness program for the MAU.

Meanwhile, MSSG 24 was fully occupied supporting the MAU. In addition, the MSSG was faced with an increased number of people and units it had to support because of the attachment to the MAU of such varied elements and equipment as an intelligence section augmentation, a large radio battalion detachment, an Armed Forces Radio and Television Services detachment, an Ashore Mobile Contingency Communications Central—a van requested and received by the 22d MAU in April—ANGLICO, and a seven-man translator team, all of which increased the MAU's strength by 170 Marines and sailors. These disparate detachments presented "a unique challenge to the MAU logistics effort due to the varied supply requirements of each and the lack of [organic] supply and maintenance support with the MAU Service Support Group. Most support is provided from external sources, such as naval supply and parent commands."[9]

On 17 June, as though anticipating future events, all MAU elements including the headquarters, took part in a mass casualty and evacuation drill ordered and directed by Commodore France.[10]

The political-military-diplomatic ferment going on in Lebanon caused Colonel Geraghty in June and early July to begin planning for a possible expansion of Marine operations to the south. The Lebanese Armed Forces had been planning to control certain areas when and if they were vacated by the Israelis. Asked in May if there was a possibility that his MAU would be expanded to brigade-size, Colonel Geraghty replied that he had heard rumors of the sort, but nothing substantive. He was prepared for such an eventuality, however.[11]

Early July remained quiet for the Marines, for the most part. Then, on 22 July, the airport was attacked by guns and rocket launchers fired by members of Walid Jumblatt's predominantly Druze Progressive Socialist Party (PSP). Nearly a dozen 122mm rockets and 102mm mortar shells exploded inside the Marine perimeter. A Lebanese civilian was killed, seven civilians and three LAF soldiers were wounded, and an American sailor and two Marines were slightly injured by shell fragments and flying glass. In the middle of the month, Marine patrols in Hay es Salaam were increasingly harassed by Lebanese civilians. One struck a Marine in the chest in a show of bravado. The Marines showed restraint and did not respond. Before long, such attacks ended.

Anti-American sentiment escalated at the end of the month, however, when a group of two or three gunmen, later identified as Shia Muslim supporters of Amal, fired short bursts from semi-automatic weapons through the airport fence at a group of Marines jogging on the airport perimeter road. No one was hit. It was learned later that this attack was meant to be a warning for the Marines not to get involved with LAF operations.[12] As a result of these attacks, the 24th MAU began to plan a series of escalated responses, ranging from non-lethal to lethal.[13]

Despite these warnings, the Marines continued unit and individual crosstraining with LAF and MNF units. This included vertical assault training and a combined amphibious landing on Green Beach with the French, and rappelling and parachute training with all allied units. In addition, Marine staff noncommissioned and noncommissioned officers filled drill instructor billets in the Army Mobile Training Team school. During all of this, visitors still arrived.*

On 18 July, Lieutenant Colonel Harold W. Slacum relieved Lieutenant Colonel Beebe as the MAU executive officer. The latter was returning to the States to take command of a squadron.

To many of those on the scene, the shelling on 22 July marked the point in time when the Marine situation in Lebanon began to deteriorate markedly. During the first three days of August, however, while it was relatively quiet, Lieutenant Colonel Gerlach rotated his line companies to relieve the boredom of remaining in one place too long, and to keep them on their toes in a new environment. Companies A and C covered the eastern perimeter of the MAU line, while Company B was repositioned to the Lebanese University. A platoon from Company C stood guard at the British Embassy and the Durrafourd Building, and a squad from the BLT's antitank platoon was in place

*In the first half of July, the MAU was visited by: General Sir John Stanier, Chief of Staff of the British Army; Vice Admiral M. Staser Holcomb, Deputy CinCUSNavEur; Vice Admiral Edward H. Martin, commander of the Sixth Fleet, who had relieved Admiral Rowden; Major General Keith A. Smith, commanding general of the 2d Marine Aircraft Wing, whose son was a Marine captain serving as the BLT 1/8 air liaison officer; Brigadier General Robert J. Winglass, commanding general of the 2d Force Service Support Group; and VIP flights carrying in Secretary of State Shultz; General Vessey; Marine Brigadier General Ernest T. Cook, Jr., Deputy Director, J-3, EUCOM; and Air Force Brigadier General Edward J. Heinz, J-2, EUCOM. In addition to these visitors, Colonel Geraghty also hosted and briefed Congresswoman Beverly B. Byron, and Congressmen Nicholas Mavroules, W. Henson Moore, and Thomas J. Huckaby. Another visitor to the MAU, no stranger to Beirut, was Colonel Robert B. Johnston, who now commanded the 8th Marines at Camp Lejeune.

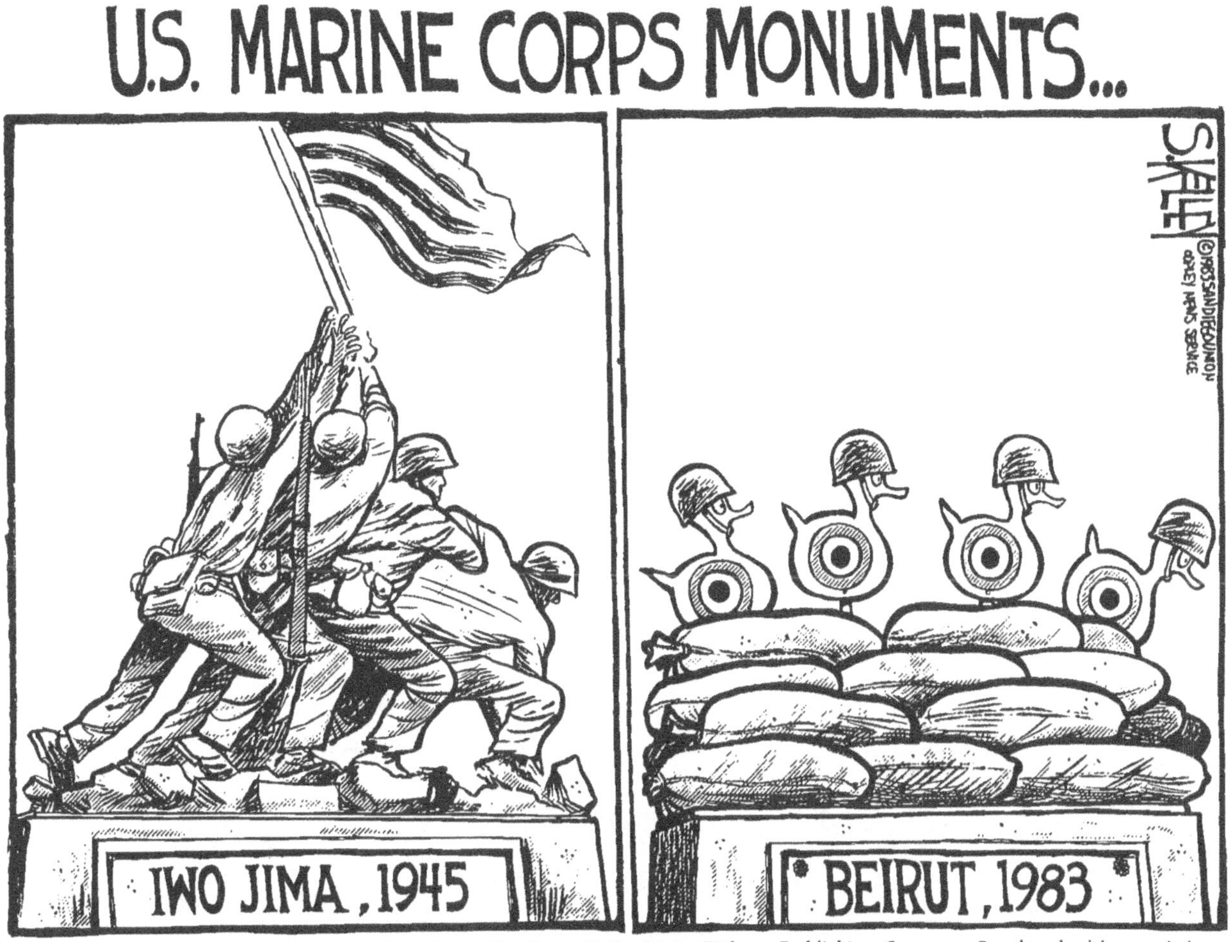

Copyright 1983. Steve Kelly. *Union-Tribune Publishing Company.* Reprinted with permission.

at the American ambassador's residence. Meanwhile, Colonel Geraghty had attached one ANGLICO team each to the Italian and the British contingents.

To better pinpoint the source of incoming artillery and rocket fire, two Army Field Artillery Schools Target Acquisition Batteries (FASTAB) equipped with AN/TPQ-36 counter-mortar/counter-artillery radars were sent to Beirut and attached to the MAU. The addition of the FASTAB provided a significant target acquisition capability that the Marines did not have earlier.[14]

These radars were put to use two days after their arrival. On the night of 8 August, two more rockets impacted at the airport, as if to announce the beginning of renewed attacks on the Marines. At 0525 on the 10th, a 122mm rocket landed in between the MAU and the BLT headquarters, slightly wounding one Marine officer.

An hour later, the MAU headquarters area came under a rocket barrage aimed at the Lebanese Air Force flight line and LAF camps immediately north of the airport. (This closed down airport operations from 10 to 16 August.) Within another hour, 27 122mm rockets exploded around the Marine positions. By 0725, having located the launching area, the BLT's 81mm mortar platoon fired four illumination rounds over the suspected sites. This told the Lebanese that the Marines knew where they were, and effectively silenced them in 10 minutes. This marked the first time the Marines had been forced to shoot indirect fire weapons in their own defense.

At the same time the rockets were landing, two Marine Cobra helicopters were launched over Green Beach, ready to attack targets on call. Simultaneously, Commodore France placed all his ships at general quarters and positioned his naval gunfire ships so they were also ready to fire on ashore targets when called upon. The 24th MAU command chronology for this period wryly notes that "Marines and sailors ashore got first-hand appreciation of things that have gone bump in the night for the past year."[15]

In late July, Walid Jumblatt, the Druze leader, had formed a Syrian-backed National Agreement or National Coalition Front which was opposed to the 17

Photo courtesy of Claude Salhani

Marines from 24th MAU parachute from a CH-53E helicopter over Beirut on 13 July 1983 in a joint training exercise with French and Lebanese troops.

May agreement between Lebanon and Israel. In anticipation of the Israeli withdrawal from the Alayh and Chouf districts, fighting escalated between the Druze and the Christian Phalangists, and also between the Druze and the Lebanese Armed Forces, which were trying to strengthen their respective positions in view of the anticipated Israeli withdrawal. The LAF was also clashing with the pro-Khomeini Amal militia in the western and southern suburbs of Beirut.

A number of rounds were impacting accidently and on purpose in the airport area because of this new fighting. Some of the spill-over fire landed on Rock Base, the Marine squadron's terminal at the north end of the airport's northeast-southwest runway. During the rocket and mortar fire on 10 and 11 August, one Marine was wounded.

As the fighting in the hills and within the city built up during the middle two weeks of August, isolated rounds landed in Marine positions and near the Ambassador's residence. Meanwhile, Marine observation posts reported sightings of the Israeli Defense Force

Copyright 1983. Vern Thompson. *The Lawton Constitution and Press.* Reprinted with permission.

"WONDER HOW MUCH LONGER WE'LL BE HERE "KEEPING THE PEACE"?"

The new Commandant of the Marine Corps, Gen Paul X. Kelley, briefs the press during his August 1983 visit to the 24th MAU. To his left is Sixth Fleet commander VAdm Edward H. Martin, USN. Gen Kelley was to return to Lebanon less than two months later.

redeploying non-essential equipment, while preparing to withdraw its combat units as the Marines continued patrolling amid minor incidents of harassment. The Marines were not the only targets of various Lebanese factions during this period; the French, British, and Italian contingents also took small arms and mortar fire.

Despite this combat activity, some crosstraining continued, liberty parties and port visits continued, and official visitors continued to arrive. Admiral Martin and General Vessey visited Colonel Geraghty on 1 August; Senator Robert W. Kasten, Jr., appeared on the 9th, followed nine days later by Congressmen Clarence D. Long, Lawrence Coughlin, William Lehman, Marty Russo, John E. Porter, and Richard H. Lehman. The highlight of the month was the two-day visit, 16-17 August, of the new Commandant of the Marine Corps, General Paul X. Kelley, and the new Sergeant Major of the Marine Corps, Sergeant Major Robert E. Cleary. On the 20th and 21st, Secretary of the Navy John F. Lehman, Jr., escorted by Admiral Martin, visited the Marines ashore and the Phibron afloat. As usual, the "Cammie Cab Company" kept busy, flying a total of 1111.6 hours in August, and carrying 90,550 pounds of cargo and 322 passengers—including Special Envoy Robert C. McFarlane, a regular customer.

The fighting in Beirut continued to escalate throughout the month, erupting with a roar on the afternoon of 28 August with heavy fighting between the LAF and the militia that continued for four days. The fighting was particularly intense around the airport, reaching its peak between 1400 on the 28th and 1230 the next day. As 24th MAU staff historian Captain Timothy J. Tanner wrote in the MAU's command chronology for August, "The small arms fire was as great as that on a 200-yard rapid fire string of the Marine Corps qualification course."[16] He also noted that in this two-day period, over 100 rounds of 82mm mortar and 122mm rocket fire landed in the airport area, with the shells landing as close as one kilometer in front of Marine positions. Two Marine-Lebanese checkpoints also received fire, as did the Marine company at Lebanese University.

In accordance with their rules of engagement, the Marines responded to these attacks with carefully orchestrated return fire. "They were [acting] in self defense, did not initiate the exchange, and ceased firing when the attackers' fire was no longer directed at them."[17]

On 28 and 29 August, it was easy to distinguish these deliberate attacks from the stray rounds that passed overhead. As the fighting increased around the airport on the morning of the 29th, Druze mortar fire continued to land on the Marine lines. The BLT's 81mm mortars fired six illumination rounds over one of the suspected firing positions at 0940 in an attempt to suppress the Druze fire. Minutes later, several 82mm rounds landed on a Company A rifle platoon position, killing one Marine and wounding four others, one of whom later died. Before the day was ended, eight more Marines were wounded.

Firing continued all morning. At 1000, Druze rockets began hitting a LAF position outside the northern portion of the perimeter. At the same time, the Marines learned that a Druze artillery or rocket battery was preparing to fire on them. The 81mm mortar platoon once again fired six illumination rounds over the suspected Druze rocket battery, which was now firing at the rate of a rocket every 15 seconds. At approximately 1150, the guided missile cruiser *Belknap* (CG 26) fired two illumination rounds from its 5-inch gun. When this didn't stop the Druze, the Marine artillery fired in anger for the first time. A new 155mm howitzer of Battery C, 1st Battalion, 10th Marines fired six 155mm, high-explosive, point-detonating rounds with pinpoint accuracy on the position, reportedly killing three and wounding 15 Druze. The Druze position went silent.

The MAU used all available resources to identify and precisely locate sites of the weapons firing at the Marines—the integrated observation station; visual sighting from the roof of the BLT headquarters building; sighting reports from observers on the lines, using the "flash-bang" ranging method; aerial sightings by observers overhead in a UH-1N and two AH-1T Cobras; and the electronic imagery registered on the screens of the FASTAB radars. This compilation of data enabled the MAU to fire with complete assurance of definite results. Colonel Geraghty also noted that he received overtures for a ceasefire beginning about 1230, which resulted in the end of hostile fire 45 minutes later. "The howitzer battery certainly reached out and touched someone. . . ."[18]

Earlier that morning, about 1045, an unidentified armored personnel carrier had opened fire on a joint-

A Battery C, 10th Marines M-198 155mm howitzer at the north end of Beirut International Airport points towards the Shiite guns which fired on Marines on 29 August.
Photo courtesy of Claude Salhani

Sign fronting Company A, BLT 1/8 tent in July 1983.

Sign in front of same mess tent, August 1983.

Same tent, same sign, new message, the next month.
USMC Photos by SSgt Robert E. Kline

ly manned Marine-LAF checkpoint with .50 caliber and 7.62mm machine guns. Two Cobras were detailed to locate the source of the fire. When the wingman saw tracers directed at the lead Cobra, he lined up on the target and fired a 5-inch Zuni rocket, silencing the machine gun. Nevertheless, the first Cobra had sustained three hits and had to return to the *Iwo Jima* for an emergency landing.

After Colonel Geraghty set Alert Condition 3 at 1745 on 29 August,[19] Lieutenant Colonel Gerlach sent an armed supply convoy out to the Marine checkpoints on the eastern perimeter. The convoy reached its destination without incident, but on the return through the village of Hay es Salaam, the Marines picked up an escort of roughly 150 masked and heavily armed civilians who guided them to the village outskirts.

The last three days of August were marked by sporadic and occasionally heavy fire fights and artillery shelling in Beirut. At the same time, Marine positions came under random fire from weapons of all calibers. Heavy fighting resumed after dawn of 30 August, when two LAF brigades attempted to sweep west Beirut clean of the Shia militia, and the firing came uncomfortably close to the US/UK Embassy. Late that afternoon, Colonel Geraghty was tasked with providing additional security for the American ambassador's residence at Yarze, and he sent an additional squad from BLT 1/8's antitank platoon by helicopter. At this time all Marine positions on the perimeter and at the university were now under attack. The Marines returned fire to the extent permitted by their rules of engagement. The ANGLICO teams soon located and

Copyright 1983. Jack McLeod. *The Army Times Publishing Company.* Reprinted with permission.

identified two artillery positions that had been firing on the French headquarters: "Discussion with the French indicated that they were not requesting that we return fire, but would not oppose it."[20] The howitzers fired two illumination bursts over each of the emplacements, quietening them for about one hour. It was later determined that one of the positions was an LAF battery firing at an Amal position located near the French headquarters, and the other was an Amal artillery battery.

On the morning of 31 August, a JCS order came down through to the Sixth Fleet, directing Commodore France and Colonel Geraghty to coordinate their efforts in drawing 500,000 rounds of 5.56mm ammunition from the MAU's contingency supply and delivering it to the Lebanese Armed Forces at the Juniyah Naval Base. The ammunition was transported by Phibron landing craft and HMM-162 helicopters. All the while, a Phalange gunboat laying off Juniyah observed the operation.

That afternoon, the LAF began shelling Jumblatt's PSP artillery and mortar positions in the hills east of the airport. Responding to this shelling, Jumblatt's guns fired on the Ministry of Defense, endangering the lives of the U.S. Army training team members there. Two Marine 155mm howitzers then fired high-explosive, variable time-fuzed shells at the PSP positions, and the shelling stopped. Gradually the fighting in Beirut tapered off, and the Marines then resumed Alert Condition 3.

The control of supporting arms during the last three days of August was divided between sea and shore. Colonel Geraghty controlled the artillery and mortars ashore, while Commodore France retained control of the naval gunfire, and control of all fixed and rotary wing aircraft. The control of direct fire weapons remained with Lieutenant Colonel Gerlach, the BLT commander.

Under the best of conditions, it was difficult to determine which of the Lebanese factions was firing at whom. As the MAU recorded in August:

> During this period, the reports were made of LAF units to the northeast of the city firing east, south, and west into the city; PSP units in the city firing east and north; PSP units in the hills firing into the city, north of Juniyah, and south; LF (Phalange) units firing to the east, south, and west into the city; and LF units in Juniyah firing south at the hills,

southeast towards the PSP positions and southwest into the city. Isolated mortars and guns were everywhere, firing in all directions. The Marines only fired east and southeast. The fire support situation was best described by the American Ambassador as being unclear as to who was doing what to whom, and why. The when was evident throughout the period.[21]

Earlier in the deployment, Colonel Geraghty and his S-4, Major Robert S. Melton, had discussed the possibility of moving the Marine amphibious bulk fuel system, which was part of the MAU deployment package, to Green Beach. They foresaw the necessity for the system in July when the *Iwo Jima* was leaving Beirut for a port visit. At the same time, a sufficient fuel supply would be needed ashore to ensure continuous helicopter operations.

The CH-53Es, each of which carried a 500-gallon bladder externally, transported the 20,000 gallons of JP-5 aviation fuel ashore.[22] The MAU also transferred 25,000 gallons of DMF (diesel marine fuel) ashore, a step that would prove to have been prescient when the fighting reached its height in late August and commercial sources in Beirut were no longer available. Commodore France, together with Major Douglas C. Redlich (MSSG 24 commander) and Major Melton, devised a plan to transfer fuel ashore without having to move shipping closer to the beach and hostile fire. Simply, they put fuel bladders on one section of the Green Beach causeway, towed it out to a ship, far offshore, and filled the bladders:

". . . and then under cover of darkness and in the emission control state, without having any communications, [we] would move that ashore to the bulk fuel system we had set up on the beach in late August, [and] with a minimum amount of equipment and a minimum amount of assets being involved, we were also able to bring ashore 20,000 gallons of fuel on two different occasions during hostile periods without ever having to threaten any high value ships."[23]

By the time the 24th MAU arrived in Beirut, the 22d MAU had already established a relationship with commercial vendors in Beirut. The Marines recognized that Beirut had sources that could provide much of what the MAUs needed. This meant:

. . . getting all our fuel from commercial sources. . . fresh fruits and vegtables, paper products for the mess halls, soda beer . . . things like that . . . and as the situation developed into August and September, there was a dramatic change in all that because obviously the door to Beirut [had] shut as of the 28th of August and all that turned around.[24]

Fuel was a critical item. It was needed to keep the helicopters flying, the vehicles rolling, and the diesel generators operating and also for use in field sanitation. As Major Melton stated, ". . . we used about 1,500 gallons of diesel fuel a day, 800 of which, amazingly, were used to maintain sanitary conditions . . . [for] the field heads. The field latrines were burned every day using five to six gallons of diesel fuel for each head."[25]

In the wake of events during 28-31 August, EUCOM suspended the requirement for the Marines' presence patrols in Beirut, and began preparing contingency plans for reinforcing the MAU. The nuclear carrier *Eisenhower* (CVN 69) and its carrier battle group, together with the French carrier *Foch* and several Italian gunfire support ships, moved closer to the Lebanese coast. With American Embassy personnel, Colonel Geraghty reviewed and updated his plans for evacuating civilians, "non-combatants."[26]

Meanwhile, on 4 September, the Israeli Defense Force began redeploying its troops from the Chouf and Alayh districts to the Awwali River in southern Lebanon, without notifying the Lebanese government, the Multi-National Force, or any of the embassies. At this time, the Lebanese Army was no more able to fill the vacuum left by the Israeli withdrawal than it had been on 17 May, when the Israeli-Lebanese Agreement was signed. Instead, the LAF moved to Khaldah, south of the airport near the Company C position. Most military observers knew that before long the LAF was going to have to clear the dissident elements from the suburbs of Beirut:

. . . but the Lebanese were not quite strong enough at that point to really dominate the areas. They could control the periphery, but they couldn't get within the city and enter Hay es Salaam, Burj al Barajinah, with those other Druze and Amal camps in order to clear them out. They could only threaten them from the outside. So, they didn't really clean it out. Subsequently, after the Israelis' withdrawal—it was not an unexpected withdrawal—but it occurred at such a time that the Lebanese had not been able to forge an accommodation of sorts with either the Druze or the Amal. Therefore, open warfare was a foregone conclusion.[27]

As the Lebanese Army began assembling its troops, the airport (Company C's position in particular) came under fire, which increased proportionally with the growing number of Lebanese troops massing and intermingling with Marines. The Marines returned fire at 11 identifiable targets, firing small arms, machine guns, and five rounds from the main gun of one of the five tanks.[28]

The withdrawal proceeded without incident for the Israelis, but it was marked by an outbreak of clashes between the Druze and the Phalange in the mountains, particularly in the areas of Suq al Gharb and Aytat, where a vacuum had begun to develop.

The fighting escalated on the 4th. Near noontime,

USMC Photo by SSgt Robert E. Kline

The 1st Platoon, Company C, BLT 1/8, manned Combat Outpost 35, "The Alamo," north of Ash Shuwayfat in the eastern section of the 24th MAU airport perimeter.

PSP units moved a 106mm recoilless rifle into position and began firing at LAF armored vehicles and Marine bunkers. Adding to this fire were mortar rounds, which landed all over the southern portion of the airport and as far north as the southern part of Green Beach. On board the *Austin*, Marine Cobras were placed on a five-minute alert, and BLT 1/8 moved a tank into Company C's lines to take the PSP recoilless rifle under fire. Meanwhile, the LAF moved a column of more than 60 mechanized vehicles south along the coast road and attacked into Khaldah behind an artillery barrage aimed at PSP positions in the foothills.

Firing at the airport then gradually tapered off, ending at 1426. Beirut International Airport remained quiet until 2000, when four rockets landed at the southern end of the Marine perimeter. This began a night-long, sporadic shelling of the airport and Green Beach. Marines fired 81mm and 155mm illumination rounds in an attempt to quiet the PSP gun, repeating this fire after 0400 on 5 September at the same targets. Two Marines were slightly wounded by PSP shell fragments, treated, and not evacuated.

Marine positions at the southern end of the airport continued to take artillery, rocket, and small arms fire as the Lebanese Army continued to use that area as a staging base. The MAU had three more Marine casualties, one of whom was hit in the neck and evacuated by helicopter to the *Iwo Jima* for treatment. Marine equipment very soon began to show signs of battle damage. "In fact, the tentage in Company C area looked more like camouflaged netting than it did tents."[29]

As the Lebanese fighting carried on into 5 September, both Colonel Geraghty's command and the Phibron alternated between Alert Conditions 1 and 2. Once again, Commodore France and Colonel Geraghty were called upon to provide a major resupply of artillery ammunition from the MAU's contingency supply and deliver it to the Lebanese Armed Forces at Juniyah. "This operation later proved to be the mainstay of the Lebanese government as the LAF, at

USMC Photo by SSgt Robert E. Kline

Marines of Company C, BLT 1/8, 24th Marine Amphibious Unit, conduct a foot patrol along the railroad tracks southeast of the company positions in late August 1983.

one point, were firing over 2,000 rounds of artillery ammunition in each 24 hour period."[30]

The next two days proved to be a particularly trying period for the Marines, both on the airport perimeter and in the area immediately surrounding the MAU and BLT headquarters. The first heavy rocket barrage began shortly after midnight, 5 September, with 11 rockets hitting near the terminal. Between 0345 and 0530, 21 rockets impacted, killing two Marines and wounding two others. Initially, the Marines replied by firing 155mm illumination rounds. With the light of day, two Cobras were launched in an attempt to discover the source of the rocket firing, but the great number of potential launching sites in the hills made it impossible to uncover the right one.

Meanwhile, the LAF attack in the hills was not going well. Lebanese government troops were forced to move east to Suq al Gharb to link up with the other government units in this strategic town, thus conceding to the Druze all high ground overlooking both the city of Beirut and Marine positions in the airport. By 1600 on 6 September, more than 120 rounds of artillery, mortar, and rocket fire had exploded at the airport, wounding another Marine.

During 6 September, Presidential Envoy McFarlane was involved in earnest diplomatic discussions with the protagonists, and even traveled to Damascus, Syria, in an unsuccessful attempt to negotiate a ceasefire. The night of 6-7 September passed relatively uneventfully, although shells continued to fall about the Marines, and the French suffered one soldier killed and three wounded when their compound came under fire. At 1815 on 7 September, three rounds landed within the Marine perimeter and several hit just outside of Company C lines. Six more shells landed in the vicinity of Rock Base, the HMM-162 landing area at the northern part of the airfield, and the MAU headquarters, wounding a Marine from the MSSG. The Marines then returned fire with six high explosive 155mm rounds on a suspected Druze fire direction center. The

USMC Photo by SSgt Robert E. Kline

24th MAU Marines set up this static display of "stray" rounds which fell on a Company C, BLT 1/8 area following heavy firing on American positions in September 1983.

Artillery fire from an unknown source hit this Company C, BLT 1/8 tent in September.

USMC Photo by SSgt Robert E. Kline

Marines on the roof of a Lebanese University building view an artillery round impacting near a Company A, BLT 1/8, position on the MAU perimeter in late September 1983.

The carrier Independence, *hull down below the horizon in the waters off Beirut, maintains her position, ready to provide air support to Marines ashore upon call when needed.*

USMC Photo

USMC Photo by SSgt Randy Gaddo

In a more peaceful moment, Marines with 2d Platoon, Company C, BLT 1/8, relax with a game of basketball at their well-sandbagged position on the perimeter of the Beirut International Airport.

artillerymen next prepared to fire a high explosive illuminating mission coordinated with naval gunfire from the frigate *Bowen* (FF 1029). The Druze position ceased firing, however, and the mission was cancelled.

French aircraft from the *Foch* flew two photo-reconnaissance missions on 7 September over the area containing the artillery emplacements suspected of having fired on the French compound. These flights were followed by an F-14 mission flown from the *Eisenhower*, marking the first use of a Tactical Aerial Reconnaissance Pod System (TARPS) mission by U.S. forces in Lebanon.[31] These TARPS missions quickly became a source of good information, used to good effect both by the MAU and the carrier battle group. Although there were Russian SA-7 surface-to-air (SAM) missiles in the hills ringing Beirut, the American command believed that their employment was tightly controlled and that they did not pose a significant threat to the F-14 flights.

Generals Miller (CG, FMFLant) and Gray (CG, 2d Marine Division) visited the MAU for three days beginning 7 September. At about 1130 on 8 September, three rocket rounds landed approximately 200 meters from where the Marine commanders were standing. In reply, a coordinated 155mm howitzer volley and 5-inch salvo from the *Bowen* landed on the target, marking the first time that naval gunfire was actually employed in support of the Marines ashore.[32] And it changed the MAU mission a bit more from one of peace-keeping presence to one of active participation.

On the morning of 9 September, 20 mortar rounds exploded near the airport terminal area, marking the first attack from a lone mortar position located southwest of the airport. For the next month, this position was to plague the Marines, who named the gunner "Ali," and "Achmed, the Mad Mortarman." More explicit expletives sometimes flew in the direction of the Druze gunner, who would fire 10 to 20 rounds and then disappear for the rest of the day. He avoided all counterbattery detection by varying his attacks in time and intensity. "Initially, the MAU referred the attacker to the LAF and watched in vain over several days as the LAF artillery landed everywhere but in the vicinity of the mortar position."[33] Most of these Druze mortar rounds were directed at the LAF Air Force's Hawker-Hunter jet fighters west of the main MAU area.

The 24th MAU command chronology for September noted that the movement of Government of Lebanon troops into the mountains put Marines into a position of providing more direct support to the Lebanese. As the LAF became engaged in the mountains along a line from Alayh in the north, and Bshamun, the firing into the Marine lines diminished and gradually ceased. For the next 15 days, the fighting in the hills above Beirut and the airport seesawed, occasionally bringing rounds into the airport area, but not in the volume of the previous two weeks.

In the first week of September, Colonel Geraghty noted, "All ops in this report were protective in nature, either passive, building or reinforcing positions, or active; locating hostile weapons firing on the BIA. Marines returned the fire where appropriate."[34] He also commented:

> The increasing involvement in direct and more frequent combat actions has tasked the MAU assets to their fullest. All hands are at quick step and the forced march pace is beginning to tell . . . . 24 MAU has added a new page to the discussion on maneuver warfare, i.e., stakes are being raised weekly and our contribution to peace in Lebanon since 22 July stands at 4 killed and 28 wounded. Phibron-8 also

added 1 wounded. We still stand our ground, however, and accomplish the mission we were sent here to do. Morale is high and while many of the Marines do not fully understand the complexities of the effort, all realize its importance to the nation. The call from the President and the visits of LtGen Miller and MGen Gray, were well appreciated and provided a needed boost at a rough time. 24th MAU will hang tough.[35]

General Mead spoke of Geraghty's problems in a presentation at the Marine Corps Historical Center in Washington, D.C., on 14 September 1983:

> . . . with the situation that you find yourself in now, what options do you have? Withdraw? Attack? Hunker down? . . . Do you attack? It's not a military problem. Who do you attack? Do you take on the Druze, the Shia, the Muslims? Who do you attack?
>
> Do you hunker down? Isn't that a wonderful expression? Hunker down. Well, you remain on the defensive right now, being responsive to the political arena in hopes that some type of political solution can be arranged through Special Envoy Bud McFarlane. . . .[36]

General Mead also referred to a telephone call from the President to Colonel Geraghty, in which the Commander-in-Chief reminded the MAU commander that he had the full support of the nation and further reminded him that he had the Sixth Fleet in direct support, and to use it when it became necessary.[37]

Despite the shift of the fighting into the hills, the Marines were not entirely forgotten by the Druze or

Photo courtesy of Claude Salhani

When LtGen John H. Miller, Commanding General, Fleet Marine Force, Atlantic, visited 24th MAU headquarters on 9 September 1983, Marine positions were hit by three enemy rocket attacks, which were answered by naval gunfire from American ships.

On the alert, Marine machine gunners survey the scene in front of their positions.

Photo courtesy of Francoise de Mulder

any of the other factions. At 0100, 11 September, Amal forces attacked a joint Marine-Lebanese checkpoint in an attempt to capture ammunition. The attack was beaten off; the Amal were unsuccessful. There were no American or Lebanese Armed Forces casualties.

The next four days, the Marines received a varied amount of small arms, rockets, artillery, mortar, and rocket-propelled grenade (RPG) fire, suffering five wounded. All line companies and manned checkpoints were fired upon and were forced to remain in a Condition 1 state of alert.

In anticipation of his Marines reinforcing the 24th MAU, Colonel James H. R. Curd, commander of the 31st Marine Amphibious Unit, and key staff officers came ashore on 14 September for a briefing of the situation by the staffs of Colonel Geraghty and Commodore France. The 31st MAU had transited the Suez Canal after a training operation in Kenya and had arrived off Beirut on the 12th. Colonel Curd's command consisted of BLT 3/3, HMM-165, and MSSG 31, and was embarked aboard Amphibious Squadron 1 shipping—the *Tarawa* (LHA 1), the *Duluth* (LPD 6), and the *Frederick County* (LST 1184). Acting as the afloat reinforcement for the 24th MAU, Colonel Curd and the Phibron 1 team joined in the planning for contingency operations, a noncombatant evacuation operation, and prepared themselves to conduct any other mission assigned to them.[38]

The 24th MAU and Phibron 8 were given additional muscle later in the month, when the battleship *New Jersey* (BB 62) arrived on 25 September to add to the naval gunfire support already available to Marines ashore.

Commenting on the arrival of the 31st MAU and the nearness of the carrier *Eisenhower* and its accompanying battle group, Colonel Geraghty said:

> The presence of ARG Alpha offshore and the overflights from the *Ike* are reassuring and well received by the Marines on the line. I am convinced that the presence of the fixed wing [aircraft] from all the members of the MNF has given those elements firing on BIA cause for concern and they have curtailed their bombardments accordingly. Morale remains high, but the bunkers are getting deeper.[39]

Earlier, on 14 September, the MAU was directed to dig again into its LFORM to provide the LAF once more with an emergency resupply of ammunition. During September, HMM-162 helicopters transported 2,424,081 pounds of cargo (including 1,345,050 pounds of ammunition), most of it for the Lebanese Armed Forces.[40]

During the night of 16 September, the Lebanese Ministry of Defense and the American ambassador's residence were shelled heavily. The frigate *Bowen* and the destroyer *John Rodgers* (DD 983) fired six naval gunfire missions, expending 72 rounds on six targets and silencing the attackers.

Colonel Geraghty and his staff soon perceived that the LAF would have to retain positions on the Suq al Gharb ridgeline for its offensive to be successful. Walid Jamblatt must have shared this view, for his Druze PSP elements began to focus their main efforts on retaking the ridge.

Operating in support of the PSP militia, Palestinian units in the mountains attacked the Lebanese government forces at Suq al Gharb early on 19 September. The fighting soon got so heavy that gunfire of all calibers could be heard by Marines at the airport throughout the morning. As the Lebanese Army's artillery stocks became dangerously low, the Ministry of Defense, through Army Brigadier General Carl W. Stiner, Ambassador McFarlane's JCS liaison officer in Beirut, requested U.S. naval gunfire support of the LAF. The Ministry of Defense reported that the Palestinians had mounted a two-battalion tank/infantry attack preceded by a heavy artillery preparation, and the LAF was in danger of breaking under the pressure of the attack.

In response to the Lebanese request, the guided missile nuclear cruiser *Virginia* (CGN 38), the *John Rodgers*, the *Bowen*, and the destroyer *Radford* (DD 968) fired 360 5-inch shells in support of the LAF over a five-hour period. The Lebanese government later reported that the Palestinians broke and ran under this devastating barrage which turned the tide of battle.

Earlier training and material support of the LAF notwithstanding, this specific instance of combat support evidently ended the perception of the Marines as neutral in the eyes of anti-government factions. As the 24th MAU executive officer, Lieutenant Colonel Slacum, later commented:

> When we provided fire on their [LAF] behalf, it did stop the attack, they were able to hold, and it did provide them a day or so of relief while they regained their composure and reinforced up there . . . .
>
> It would appear that our very presence, even before our active support of the LAF, was having a great impact on the issue within Lebanon. It also became intuitively obvious that while we were very cautious in our exchange of artillery and naval gunfire with those batteries that were shooting at us, we did so in a manner as to, I think, show the other side that we were using great restraint. . . . And while we weren't necessarily looked upon as either neutral or friends, it was apparent we weren't looked upon as enemies, either, that the Amal and the Druze appeared to go out of their way to ensure that they did not list us as enemies . . . by and large, they did not group us, target us as an enemy. Those factions that did finally target us appeared to be from out-

Photo courtesy of Francoise de Mulder

Filling sandbags was a never-ending chore during the 18 months Marines were in Beirut. In that period, Marine Amphibious Unit troops filled more than one million bags.

side Lebanon, instigated by other nations for whatever purpose, ultimately to discredit us.[41]

Colonel Geraghty recognized that providing U.S. naval gunfire support to the LAF had changed the nature of his mission. The Marines were now considered legitimate targets by anti-government forces. Commenting on this matter in an interview conducted during his return home from Beirut in November 1983, he remarked:

> The firing we did in support of the LAF up at Suq al Gharb, that clearly changed our roles . . . . It's a milestone, no question about it in my opinion. It moved us from a previous, very careful, razor edge line of neutrality that we were walking, and treating all the Lebanese communities alike . . . . When we provided support . . . [to] the Lebanese up in Suq al Gharb, that, to me, moved it to a different category. . . .
>
> The Lebanese had run quite low on ammunition and it would have been unconscionable for us not to have provided support at a very crucial time for them. . . .[42]

For the next several days, the Ambassador's residence and the Ministry of Defense came under heavy shelling, causing fires in the residence. All embassy personnel except the Marine guard and radio watch were moved into the Presidential Palace, and the *Bowen* and *Virginia* engaged the hostile firing battery with 30 5-inch rounds each.

A change began to appear in the MAU's official report:

> Naval gunfire became the weapon of choice, if it could engage the enemy firing units, as it gave some separation from USMNF and did not require them to use their organic howitzers in defense of the LAF or MOD [Ministry of Defense].[43]

This marked the first time the anti-government elements were referred to as "enemy" in any Marine report.

Another "first" occurred on 20 September, when two Navy carrier-based reconnaissance aircraft were attacked by a SA-7 surface-to-air missile. Fired from an unknown location, the missile never acquired the lead aircraft, missing it by two miles.

The next night, 21 September, the *Radford, John Rodgers* and *Virginia* fired 90 more 5-inch rounds on two targets.

On 23 September, fighting around the airport intensified once again. Two Marine checkpoints came under heavy attack, which included fire from a 20mm antiaircraft cannon. Simultaneously, both the French and Italian compounds were taken under fire. By 1325, the fighting between the LAF and Amal militia in Hay es Salaam had become particularly heavy. Two hours later, the spillover fire began to endanger the Marines and the MAU's 81mm mortars fired 12 high-explosive rounds at a suspected Amal position, silencing it. During the following hour, the Marine mortars fired an additional 28 rounds after their positions at the Lebanese University and other checkpoints once again received intense small arms and RPG fire. That evening, Marine 155mm howitzers and naval gunfire engaged an artillery position that was firing on the airport. Later, the MAU command post came under fire, generating yet another response from Marine mortars and naval gunfire.

These outbreaks of heavy fighting highlighted the vulnerability of the isolated Marine checkpoints, the inability of the 24th MAU to reinforce and resupply them when necessary, and difficulties in evacuating the wounded without placing other Marines in jeopardy. These checkpoints initially had been established as a buffer between the Israeli forces and Hay es Salaam. Since the Israelis were no longer in the area, the checkpoints no longer served a useful purpose and the risks they faced were now unacceptable. On the afternoon of 24 September, after notifying the Lebanese government, 24th MAU redeployed 15 Marines from one checkpoint and 25 from the other to Company B positions at the Lebanese University.

At this point, the shuttle diplomats managed to arrange a ceasefire to take effect on 26 September. Just before the ceasefire, there was considerable fighting around the airport, as various factions attempted to gain favorable positions, and fighting in the airport area eventually tapered off by the end of the month. Other factors contributing to a reduction in the number of attacks on the Marines were the arrival of the reinforcing 31st MAU and the battleship *New Jersey*, as well as the demonstrated readiness of the Americans to employ naval gunfire in support of the LAF and in their own defense. The right of self-defense was also being exercised by allied forces. On 23 September, the French conducted an airstrike against the weapons that had been firing on their positions.

The 24th MAU's command chronology for September notes that the month ended in a positive vein, despite two adverse events. One was the crash of a Marine Cobra helicopter into the sea.[44] Both pilots were rescued, and had but minor injuries. The second event was the seizure by the Amal militia of two soldiers from the Army Field Artillery School's Target Acquisition Battery, when the Americans made a wrong turn away from the main north-south thoroughfare in the center of Beirut. The two were brought to Nabih Berri, the leader of the Amal, ". . . who apologized for the incident."[45] Saying that the Amal held no animosity toward the Americans, Berri released the two to a French liaison officer. Their jeep was returned, but the pistol of one of them was not.

Of this event, Colonel Geraghty observed, "The incident was particularly serious as it pointed out the relative inability of the USMNF to react to incidents of this nature and demonstrated the variety of threats to the MNF and their possible consequences."[46]

Near the end of the month, on 24 and 25 September, Colonel Geraghty hosted a large congressional delegation led by Congressman Samuel S. Stratton of New York.* Included in this group were Representatives William L. Dickinson, William Nichols, Larry J. Hopkins, Bob Stump, Beverly B. Bryon, Richard B. Ray, John McK. Spratt, Jr., Solomon P. Ortiz, and

---

*Upon his return to Washington, Congressman Stratton wrote the Commandant:

Dear General Kelley;

I wanted to take the opportunity to write you concerning the truly outstanding service of one of your officers, Colonel Timothy Geraghty, Commander, 24th Marine Amphibious Unit.

As you know, I had the honor recently to lead a delegation of 10 members of the Committee on Armed Services to Lebanon to review the difficult military and political problem firsthand. During our visit, we were able to spend several hours with Colonel Geraghty and his men at Beirut International Airport.

I know I speak for all the members of the delegation in expressing nothing but the highest praise for Colonel Geraghty and, of course, the personnel of the 24th Marine Amphibious Unit. The circumstances presented by U.S. participation in the Multinational Force (MNF) involve extremely difficult exercise of judgment by Colonel Geraghty as the on-site commander balancing the safety of his men with a political requirement to minimize the level of U.S. involvement in the area. It was apparent during our visit that the U.S. participation in the MNF was contributing to stability in Lebanon. This success can be attributed in no small part to the performance of Colonel Geraghty.

The Marine Corps and the United States of America can be justly proud of the service being performed by Colonel Geraghty.

Sincerely,
/s/ Sam
Samuel S. Stratton
Head of Delegation

Congressman Samuel S. Stratton ltr to CMC, dtd 6Oct83. Handwritten at the bottom of the letter was the note, "We also are deeply grateful for the outstanding assistance General Mead gave our delegation."

USMC Photo by SSgt Robert E. Kline

Adm James D. Watkins, Chief of Naval Operations, talks to 24th MAU Marines at the MSSG building at Beirut International Airport on 5 October 1983, during his visit.

Duncan L. Hunter. Accompanying the delegation was General Mead, who was now Director of Manpower Plans and Policy at Headquarters Marine Corps.

Beginning 26 September, the 24th MAU began a series of rest and recreation tours for the line Marines on board Task Force 61 shipping. This gave them an opportunity for hot showers, hot food in a "safe" atmosphere, and just a chance to get away from Beirut for a short period.

After being closed to all air traffic for six days, Beirut International Airport reopened on 30 September, the day that Ambassador McFarlane, a retired Marine lieutenant colonel, toured the MAU's positions. His helicopter, on an aerial reconnaissance, was hit by a stray round. Despite this incident, Colonel Geraghty commented that the ceasefire had been a welcome relief and that all MAU personnel were getting some rest as a result. He also noted that tensions remained high in Beirut, and that it was still risky to travel through most of the city.[47]

At the beginning of October, the LAF began to receive additional American shipments of armored personnel carriers, tanks, and howitzers. The training of Lebanese recruits (and units) resumed. Walid Jumblatt declared the establishment of a separate governmental administration for the Chouf and called for Druze to defect from the LAF. By mid-October, he joined with other faction leaders in agreeing to conduct reconciliation talks in Geneva.

Going into October, the ceasefire that had been negotiated in late September seemed to hold, but in a very fragile way. Sporadic fighting continued in the suburbs. The 24th MAU began the month in Alert Condition 3, but spillover fire dictated a higher state of alert. Lieutenant Colonel Gerlach rotated his companies into new positions, which they would hold until relieved by the 22d MAU in November.

On 5 October, the Chief of Naval Operations, Admiral James D. Watkins, and the Sixth Fleet Commander, Admiral Martin, visited the MAU. After receiving a briefing, the CNO in turn briefed Marines, sailors, and soldiers at the BLT headquarters building.

In time, the MAU squadron's helicopters became targets of small-arms fire. Several planes were hit, but landed safely with minor damage and no crew injuries. To forestall the possibility of more serious incidents, Colonel Geraghty changed the helicopters' flight patterns and varied their routes to and from the airport.

Meanwhile, the ceasefire continued to unravel, PLO members attempted to infiltrate back into Sabra and Shatila refugee camps, and violence erupted as barricades were set up at Hay es Salaam, Burj al Barajinah, and Ash Shuwayfat. Fighting between the PLO and the LAF caused more stray rounds to land in Company A's area at the Lebanese University. Snipers began firing at Marines from buildings close to the MAU's lines, and armed militiamen were spotted entering building ruins.

On 8 October, heavy fighting erupted between the Lebanese government forces and the militia at Burj al Barajinah, Ash Shuwayfat, and Khaldah. Poorly directed fire impacted into Marine positions most of the day, and one Marine was wounded slightly. Around 0900, Marines at the university were taken under sniper fire. One Marine was hit in the shoulder, but was not wounded seriously enough to warrant evacuation. An hour later, a CH-46 drew hostile fire with two rounds striking the aircraft causing damage to its skin. These last two incidents removed any remaining doubt that the USMNF aircraft and men were once again targets of snipers.[48]

Factional clashes and sniper activity continued for the next few days, with Marine line companies being fired at and returning fire where targets were identifiable. Of this period, Colonel Geraghty said:

> The ceasefire, while holding for the most part throughout the area, has degenerated to isolated attacks upon the MNF and low level, but violent, confrontation between the various factions; the direct threat against the Marines has increased significantly as several of the more radical groups view the MNF as an alternate and readily visible source against which to demonstrate their [hostilities]. We have returned to our most alert condition and will remain at such for some time. Resupply of my outlying units continues on the ground. I feel it is not worth it to hazard the aircraft. I have moved my air operations to a more protected area and have increased the visible presence of the tank unit to be prepared for fire if necessary. I anticipate the attacks to continue, and have commenced actions to force the LAF to take action in the Hay es Salaam area. I have identified what I believe to be a major source of the attacks upon my positions, and should they continue, and the LAF does nothing, I will reduce the threat as effectively as possible.[49]

The squadron's operations center ashore, at the northern end of the airfield, was evacuated:

> . . . because we were just taking too much sniper fire. It was no longer safe to walk out in the flight line. The snipers were there at the end of the runway, had [us] zeroed in and you just couldn't walk out there without drawing fire. So, we had to close that area completely.[50]

At one point, the squadron's S-2 chief, a trained sniper, was called upon to return fire during one period of heavy sniping and he reported a kill. With the exception of a small crew that remained ashore to operate Landing Site Brown on the southwestern portion of the airport, all HMM-162 Marines went back to the *Iwo Jima*.

On 12 October, General Kelley, Commandant of the Marine Corps, paid his second visit to Beirut. The next day, he awarded 12 Purple Hearts to Marines wounded in Beirut, held a press conference, and returned to Washington.

Despite the worsening situation, the 24th MAU was able to send Marines on liberty runs to Turkey and Alexandria, Egypt. Isolated hit and run attacks against Marines continued. On the evening of 13 October, a grenade was thrown from a car speeding on the Corniche at a sentry standing guard at the Durrafourd Building. He was wounded badly enough to require evacuation to the *Iwo Jima* after initial treatment at American Hospital. At 2103, Marine helicopters at Landing Site Red at the northern end of the airport, near the Marine artillery positions, came under heavy small-arms and RPG attack. A short time later, fire from Hay es Salaam hit the northern Marine perimeter.

At 1030 the next morning, accurate sniper fire was directed at two Marine jeeps traveling the airport perimeter road in the eastern sector. The driver of the first jeep was hit in both legs. The driver of the second jeep was shot through the chest and the vehicle overturned. Marines at the landing site and on the perimeter returned the fire with unknown results. Both Marines were evacuated to the *Iwo Jima*, where the second jeep driver died of his wounds. Both LS Red and the perimeter road were subsequently closed as sniper fire became a daily hazard.

Militia activities in Hay es Salaam continued to escalate as militiamen, clearly visible to the Marines, prepared sandbagged positions in the ruined buildings opposite the Marine lines, stocking them with weapons and ammunition. The night of 14 October was particularly threatening as militia snipers fired into the positions of Companies A and C. This sniper fire continued into the morning of the 15th, and the Marines deployed a sniper team of their own to deal with this new threat. "The team surveyed the area with sniper scopes for several hours, pinpointing the snipers actually firing at Marine positions. The team then opened fire with 18 rounds of match 7.62 ammunition at 14 targets. Their success was evident by the sudden silence from each hostile position."[51] Other firing into Marine positions, however, continued to be hostile and unpredictable.

At 1615 on 16 October, a tense calm was shattered

when Company A, at the university, began receiving rifle and machine gun fire from a bunker near its position. Marine snipers returned the fire for the next four hours. At 1915, the volume of fire increased. Three Marines were wounded when five RPGs detonated near the library building. Company A returned fire with rifles and machine guns. The heavy fire kept a medevac helicopter from landing to take out the wounded. At this point, the British contingent, across the Sidon Road, offered their Ferret scout car to escort a Marine motor convoy to the Ministry of Defense, where the two most seriously wounded Marines could be flown out to the *Iwo Jima*. While flying over Hay es Salaam, the choppers were fired upon, but not hit. The hostile fire against Company A increased, and the Marines fired two Dragon antiarmor guided missiles to silence a particularly troublesome machine gun. Another Marine died when Company A's forward air controller was shot in the forehead by a sniper. The volume of fire was such that the dead Marine and two other wounded men could not be evacuated until late in the afternoon of the following day, after a convoy returned to MAU headquarters by a roundabout route.

On the night of 16-17 October, the firing spread to the north and south of the university buildings, placing Company A under siege. By midnight, the fire had begun to taper off, even though it continued sporadically from Hay es Salaam. Scattered fighting continued in Beirut and its suburbs until the 19th. At this time, Lebanese government forces were heavily engaged at Suq al Gharb and in the vicinity of Burj al Barajinah, as well as in Khaldah.

At 1622 on 19 October a supply convoy returning from the Durrafourd Building and the British Embassy was hit by a remote-controlled car bomb. Four Marines were wounded by the explosion. As the Marines rushed a covering squad to the scene of the bombing, Italian soldiers on the scene treated the wounded, and evacuated the most seriously injured to the Italian field hospital. Later, intelligence sources revealed that the Americans had been targets of a pro-Iranian Islamic fundamentalist sect. Four days later, a French jeep was bombed when it passed a command-detonated explosive device hidden in a garbage can. One French soldier was wounded.

Beirut was quiet on 20 and 21 October, with only one Marine checkpoint reporting incoming fire. Lebanese units at Suq al Gharb and Khaldah exchanged artillery fire with the Amal militia, but only Company B on the eastern perimeter received any hostile fire. On 22 October, it was quiet enough for the sailors and Marines to enjoy a USO show at the BLT building.

For obvious reasons, cross-training with the LAF and other MNF units had to be severely curtailed during October. At the same time, the 24th MAU began looking forward to its mid-November relief by the 22d MAU. Backloading of non-essential equipment began on 15 October as the MAU began washing down its

A British Ferret armored vehicle and a Land Rover at the 22d MAU hospital in late 1983.
Sketch by Maj John T. Dyer, Jr., USMCR (Ret)

USMC Photo

A touring USO band plays in front of the BLT headquarters building the day before it was destroyed.

lower priority rolling and tracked vehicles to shorten the time required for such efforts at Rota. Some equipment was even re-embarked on the *El Paso* before it left for a port visit to Antalya on 11 October.

During this period, a continuous flow of messages from Task Forces 61 and 62 carried word of the increasing threat up through the chain of command. Nevertheless, the watch officer in the National Military Command Center in Washington were unprepared for and shocked by Commodore France's flash message at midnight on 22-23 October (0700 in Lebanon):

> Explosion at BLT 1/8 Hq . . . a large explosion at BLT 1/8 Hq Bldg collapsed the roof and leveled the building. Large numbers of dead and injured. Are using MSSG 24 and Italian MNF medical and will medevac out of LS Brown . . . French report a Bldg in their sector also bombed . . . unknown injured; BLT Hq destroyed. Amplifying info to follow.[52]

As later messages flowed in detailing further damage and loss of life, the shock deepened. Colonel Geraghty had arisen at about 0530, going down one floor to his command operations center on the first floor of the MAU headquarters building. Noting that the night had been relatively quiet, the MAU commander soon returned to his second floor office/sleeping quarters.

The explosion of the truck bomb literally blew out all of the windows in the MAU headquarters. To forestall any injuries should such an attack ever occur, Colonel Geraghty and his executive officer earlier had taped all of the windows in their room. The explosion also cracked the MAU headquarters structure, scattering debris all over. Geraghty ran outside, where the atmosphere was foggy with debris floating down. He ran around to the back of the MAU building, seeing only, ". . . a heavy fog and debris . . . still floating down."[53]

The MAU commander next ran over to the aviation safety building behind the MAU headquarters, where all the windows had been blown out. "And it was just then . . . the fog was clearing, and I turned around and the BLT building, the headquarters, was gone. I can't explain to you my feelings then. It was just unbelieveable."[54]

Colonel Geraghty and the on-scene Marines immediately began to rescue the injured from the ruins, and to implement their mass casualty evacuation plan, ". . . because of the increased hostilities that had occurred as well as the changing situation, we had updated and practiced our NEOps with mass casualty evacs in the event of something like this. And that

As viewed from Marine positions in the southeast sector of the 24th MAU perimeter, smoke rises from the devasted BLT headquarters building in this photograph taken within seconds after the blast occurred.

USMC Photos by SSgt Robert E. Kline

Some views of the destruction following the bombing of the BLT headquarters building.

USMC Photo by SSgt Robert E. Kline

A view of the crater made in the first floor of the BLT headquarters building by the explosion of the truck bomb which devastated the structure on 23 October 1983. The arrow points to a crankcase, all that remained of the truck after it was detonated.

proved to be quite helpful."[55] Unfortunately, the battalion surgeon who would have played a very large role in treating the wounded, was killed in the explosion.

Lieutenant Colonel Slacum had been in his bunk at the time of the explosion. His mosquito netting was down and caught the pieces of glass blown inwards. The executive officer was dressing hurriedly when Colonel Geraghty returned and said, "Good God, the BLT is gone! You won't believe it, the building is gone."[56]

Geraghty then went to his communication section to report via secure-voice radio to the Sixth Fleet commander. He next sent an OpRep-3 Pinnacle Front Burner[57] flash message to the National Military Command Center in Washington, requesting that the BLT headquarters be replaced by the 2d Marine Division's air alert force and that he be sent an additional rifle company to enable the MAU to become operational again as soon as possible. Additional Marines would be needed to provide increased security for the grim clean-up facing the MAU.[58]

Meanwhile, Lieutenant Colonel Slacum had traced Colonel Geraghty's steps to the devastated area:

> . . . down to the steps of the BLT, and all the while I'm walking through debris that's about midcalf deep, and I just didn't even notice it. It's just one of those surrealistic scenes where things are . . . so grotesque and so odd that your mind doesn't comprehend, you're still in somewhat of a little bit of shock and I just didn't notice the stuff until I got to the steps of the building and I looked and the thing that struck me is that it was deathly silent. This was perhaps three, four minutes after the explosion, after we had run up and down the ladder a couple of times. . . . And there was a gray dust over everything you could see, as far as you could see. The concrete . . . from this collapsed building, that had once been three-four stories high was now down to one story . . . you could make out which was the first story and then just another 10-15 feet of rubble piled on top of that. . . I first looked around and that's when you started to see the first bodies, and went to check those that I could see in front of me and then realized the magnitude of the problem. I heard no one, I saw no one.[59]

Slacum then rushed to the communications office to enter the radio net connecting all of the Multi-National Forces to advise them of what had happened and to ask the Italians to send all the medical assistance they could spare. There was an immediate communications problem. The MAU headquarters did not directly monitor the tactical radio net of the line companies, which terminated at the BLT headquarters. So the MAU had to establish direct radio contact with the line companies, informing them of what had happened, and linking them directly into the MAU's combat operations center.

The MAU's operations officer, Major George S. Converse, was then in Norway making preparations for a later training exercise. His duties were assumed by the fire support officer, Captain Timothy J. Tanner, and the Assistant S-3, First Lieutenant Stephen N. Mikolaski. Among other things, they had to ensure that the fire support coordination net previously run by the BLT communication section was reestablished by the MAU. They also put the naval gunfire support ships on alert and reestablished radio communications with the FASTAB (Field Artillery School Target Acquisition Battery).

By now, rescue efforts were fully underway. Lieutenant Colonel Slacum asked the *New Jersey* to send her Marine detachment ashore to provide security; ". . . we had everybody we could get who was familiar with weapons and had been trained as a guard."[60]

Lieutenant Mikolaski, who bunked in one of the four garages in the MAU building—three doors away from the command operations center—first thought that a satchel charge had been thrown into the command post. He ran out ". . . and saw that the glass had been blown out of the door. In fact, the doors in the CP had been blown off their hinges, and everyone in the CP was either on the floor or getting up."[61]

Mikolaski found Captain Tanner in the MAU communications center first trying to raise the BLT headquarters on the radio. He then attempted to obtain reports directly from the line companies. Word was received over the MNF radio net from the French that one of their buildings had been bombed. Lieutenant Mikolaski thought for a moment that all the MNF contingents were being hit with missiles.

After ensuring that the OpRep-3 message had gone out, Mikolaski, together with MAU air officer Major Randolph P. Cotten and MAU Sergeant Major Richard E. Dudley, jumped in Cotten's jeep to see if the MSSG had been able to set up a triage station at LZ Brown. Commodore France had sent a medical team to the airport and working parties from the *Harlan County* and the *Portland* had gone to Green Beach. France had also activated medical teams on the *New Jersey* and the *Virginia* to be helilifited to the airport. He recalled the *Austin* from its port visit to Alexandria and alerted the Royal Air Force hospital at Akrotiri, Cyprus to the need for possible medical assistance. Finally, he requested medevac aircraft from Stuttgart, Germany.[62]

Cotten, Mikolaski, and Dudley then drove to the BLT headquarters building. They saw many Marines digging by hand through the ruins in a desperate attempt to rescue the living, trapped Americans. The wounded were evacuated from the MSSG headquarters, near the MAU command post, while some were taken to local hospitals in Beirut for treatment. The trio then went back to LZ Brown, where Mikolaski noted that a medical officer had arrived and the triage process was working; and these wounded were evacuated seaward.

At this moment, when working parties were frantically digging in the debris and rubble of the destroyed building with bare hands, perhaps one of the busiest Americans at the scene was Lieutenant Commander George W. Pucciarelli, the MAU's Catholic chaplain. Pucciarelli wore the wings of a parachutist earned while serving with a Marine force reconnaissance unit. Like the others in the MAU headquarters, he had been awakened by the blast. In the cot next to him was the Sixth Fleet Jewish chaplain, Lieutenant Commander

By hand, and with pick and jackhammers, rescue workers begin recovery operations at the scene of the bombing almost immediately after the explosion on 23 October 1983.

Arnold E. Resnicoff, who had arrived on Friday, 21 October, to conduct memorial services at the MSSG headquarters for the slain jeep driver. Both chaplains had run down to the BLT headquarters site. "I kept looking for the building. As I came around the edge of the shrubbery, I found out that the building wasn't there anymore," remembered Pucciarelli. "It was leveled . . . I could see the grey ash and dust just all over the place, on jeeps, on grass, on trees, on all the rubble that was down there. And then suddenly, I began to see things move within the rubble, and then I realized that these things . . . moving were our fallen comrades, were those who were wounded."[63]

The two chaplains began digging through the rubble with the others, seeking the injured and the dead. Chaplain Pucciarelli continued:

> I had my vial of oil and my stole on and I started giving last rites to the dead and seriously wounded. I remember I kept yelling for corpsmen and for assistance . . . as people were starting to come down the stairs toward the building, 'Over here, there's a man hurt over here, get a stretcher, bring him out of here,' and just going from one part to another was what I did for the whole time there. I would say that I probably saw in the first day 150, easily, wounded or dead. . . .
>
> I would stay day and night as they were pulling out the bodies, just the flash of faces that would go through your mind of . . . these guys who would talk to you, no matter what faith they were—Catholic, Protestant, Jewish. This was my second float with them and knowing some of their families and some of their kids and now realizing they were gone was just again a horrendous thought, that so many had been wiped out in one fatal blast.[64]

Within a short time, Italian and Lebanese forces joined the rescue efforts. Despite their own problems,

24th MAU Chaplain George W. Pucciarelli shows the strain of his ministrations to the living and the dead.

Photograph by Mike Lyongo, Black Star

USMC Photo by SSgt Robert E. Kline

Sixth Fleet Jewish Chaplain LCdr Arnold E. Resnicoff wears a camouflage yarmulke given him by Catholic Chaplain Pucciarelli when one he had was lost.

the French sent a contingent to the blast site, as did the British and the Lebanese Red Cross. "I kept seeing the same faces over and over again. It was remarkable how much work they did."[65] The MAU owned no heavy equipment capable of lifting the large blocks of steel-reinforced concrete to get to those who were still alive, as well as to those who were dead. Within a short time, the Lebanese construction firm of Oger Liban, which had provided such extensive assistance following the bombing of the American Embassy, arrived with large cranes and other needed heavy equipment. Lebanese civilians came to help, but some individuals were apparently there just for other purposes, e.g., looting. Chaplain Pucciarelli apprehended a number of such looters and had them thrown out of the area. Meanwhile, snipers fired intermittently at the rescue scene.

The count of casualties continued to mount. Back in the United States, Americans found that their favorite Sunday television programs were being continually interrupted by special news reports telling about the bombing. Throughout the day, satellite pictures of the bombing appeared on American television screens. Viewers could see the extent of the damage: bodies being carried out; the shocked faces of Marines as they went about their grisly business. The story dwarfed all others. Almost immediately, the country and the world were plunged into mourning.

Photo courtesy of Claude Salhani

A Marine wipes a tear from his eye as he avoids sniper fire being placed on the rescue teams searching for survivors in the bombed BLT headquarters building.

Parents, wives, friends of the 24th MAU Marines began their vigil—anxiously awaiting the casualty lists. Many old Marines, and young Marines, too, wept bitter tears of rage, frustration, and sorrow at the tremendous and senseless loss of life. Marine families in Camp Lejeune and its Jacksonville environs, and Americans throughout the country, began wondering if "presence" in Lebanon was worth the loss of a single Marine.

As rescuers continued pulling bodies from the building, they faced a major problem in identifying the dead and injured. Many of the men had removed their identification tags before going to sleep the night before. These dogtags, normally worn around the neck on chains, made sleeping uncomfortable. Many of the troops slept in their gym shorts or other athletic gear, which were not marked with their names as uniform items were required to be. Compounding the problem was the fact that all of the BLT's service record books and medical records were in the battalion administration offices in the basement of the destroyed building. Most were not recovered for several days. Some were never recovered.

The MAU began by identifying the living. Lists were made of those who had been in the building and survived. Some sort of identification was made of everyone, living or dead, who was evacuated from LZ Brown, ". . . maybe just a name, maybe a Social Secu-

Marines operate rental Lebanese heavy equipment at the bombed-out building.

USMC Photo by SSgt Robert E. Kline

USMC Photo by SSgt Robert E. Kline

Lebanese rescue workers recover a body from the ruins of the destroyed BLT headquarters.

rity number, something, but we knew how many actual people were on the bird and who they were."[66]

The first aircraft to evacuate the severely injured arrived at Beirut International Airport at about 1030 on the 23d. It was an Air Force C-9 "Florence Nightingale" from Germany and specially equipped for medevacs. The C-9 was not capable of carrying a large number of litter patients, but it had a surgical team on board. It carried out a number of the walking wounded and five litter patients.[67]

Shortly after the arrival of the C-9, a Royal Air Force C-130 arrived. It took off at 1421, headed for Akrotiri, Cyprus, with 20 more of the severely wounded Americans on board. Meanwhile, Air Force C-9s and C-141s were landing, loading, and taking off immediately for Naples, Cyprus, or Germany. The Beirut airport had been closed for several days before the explosion, but it was opened for these flights despite the risk of incoming fire.

The Phibron continued to send working parties to the bombed site from its five ships. Colonel Geraghty reported that the Oger Liban construction firm's workers and the Italian contingent were still using their heavy equipment to clear the site. He also reported that all access roads to the MAU compound and outlying companies had been barricaded with sand-filled barrels and rolling stock.

On 24 October, President Francois Mitterand of France visited Beirut to inspect the bomb damage at the French billet. He also visited the BLT building site and paid tribute to the dead Marines. Meanwhile, rescue efforts continued day and night with little sleep for the survivors.

Back in Washington, senior officials had been in motion ever since receipt of the initial flash message that announced the bombing. Shortly after being notified by the Marine Corps Command Center, the Commandant received a telephone call from the President, who was in Augusta, Georgia, telling him that he was cutting short his weekend vacation and would be returning to Washington as early that morning as possible.[68] Mr. Reagan asked General Kelley if he could join him for a meeting of the National Security Council at 0930. During this meeting, the situation in Beirut was briefed and a number of options were discussed.

It was decided that the NSC members should study these options and return later in the day for more comprehensive discussions. At this later meeting, the President appointed General Kelley as his personal representative, and directed that he go as rapidly as possible to Beirut to determine what additional security requirements might exist to protect his Marines further. As General Kelley left the White House Situation Room, the President put his arm around the general's shoulder and said "warmly and sincerely, 'Vaya Con Dios'—Go with God!"[69]

At approximately 0900 on 24 October, the Commandant's party left Andrews Air Force Base outside of Washington on board Air Force Two. The party consisted of General Kelley; Congressman John P. Murtha; Presidential Assistant Edward V. Hickey; Brigadier General Mead; Marine Colonel Matthew P. Caulfield from the White House Military Office; Lieutenant Colonel Bruce R. Greisen, head of the Counterintelligence Branch, Intelligence Division, HQMC; and Sergeant Major of the Marine Corps Robert E. Cleary.

The plane arrived at Frankfurt during the early evening. As General Kelley left the plane, an Air Force colonel advised him that a flight from Beirut bearing remains of 140 Marines who had been killed in the bombing had just arrived. General Kelley went immediately to the site, and watched as young airmen from the Air Force tenderly and respectfully removed each casket from the C-141 aircraft. While Lieutenant Colonel Frank Libutti, the Commandant's senior aide, did not accompany the Commandant at this point, he recalled that "just from talking to him later, after the fact, that it was a terribly emotional situation."[70]

Early on 25 October, while Marines were landing on Grenada, the party left Stuttgart by helicopter for the U.S. Air Force Regional Medical Center in Wiesbaden. The most poignant moment of this part of the trip came when General Kelley met Lance Corporal Jeffrey L. Nashton in the intensive care ward. Nashton was in a ". . . critical condition with more tubes going in and out of his body than I have ever seen."[71] The Commandant continued, "When he heard me say who I was, he grabbed my camouflage coat, went up to the collar and counted the stars. He squeezed my hand, and then attempted to outline words on his bedsheet. When what he was trying to write was not understood, he was given a piece of paper and pencil, and then wrote 'Semper Fi.'" General Kelley read this, his ". . . face became animated in a great combination of joy and tremendous pride all wrapped around this very heavy emotional environment. . . . This guy in a single act, in a moment, captured . . . the courage of that man and love for the Corps and his country. And more than anything, the faithfulness, the loyalty . . . the opposite of despair, you know, 'Semper Fi.'"[72]

General Kelley arrived in Beirut early Tuesday afternoon, 25 October, and was met by the new U.S. Ambassador to Lebanon and an old friend, Reginald Bartholomew, and General Richard L. Lawson, Deputy Commander in Chief of the European Command. He immediately went to the BLT headquarters to see the

Gen Paul X. Kelley, Vice President George Bush, and 24th MAU commander, Col Timothy J. Geraghty discuss the situation during recovery operations after the bombing.

extent of the damage and witness ongoing rescue operations. While there, the 24th MAU received warning of another terrorist attack, which did not materialize.

That evening, General Kelley conducted a closed meeting in *Iwo Jima's* flag mess with General Lawson; Sixth Fleet Commander Vice Admiral Martin; Mr. Hickey; Marine Brigadier General Ernest T. Cook, Jr., Deputy J-3 of EUCOM; General Mead; Commodore France; and Colonel Geraghty. Colonel Geraghty briefed the group on the current situation, and what actions he had taken to improve the defenses.

Vice President Bush arrived on board the *Iwo Jima* early the next morning, 26 October, and was briefed on the situation by Colonel Geraghty and Commodore France. Subsequently, he toured the ship's sick bay, where he presented Purple Heart medals to injured Marines and sailors.

The Vice President then visited the site of the bombing, followed by a call on President Gemayel, accompanied by Ambassador Bartholomew, General Kelley, and General Lawson. When the Vice President left Lebanon, the Commandant then called upon General Tannous, whom he had met before. Of this meeting, the Commandant's aide recalls, ". . . the meeting was characterized by the very positive interaction of the obvious dynamics between General Kelley and General Tannous . . . sort of old warriors, then trusted veterans . . . there is a charisma there that was very obvious to even the most casual observer."[73]

For the remainder of the day, General Kelley met with other MNF commanders, and visited the site of the terrorist attack against the French contingent. At approximately 1700 that evening, the Commandant called a meeting at the headquarters of the 24th MAU to discuss the contents of the report he would make to the President. This meeting included all of the principals, including Ambassador Bartholomew and General Lawson. Once the general outline and contents were agreed to, the Commandant and his party (less Congressman Murtha who had left earlier) flew out of Beirut International Airport bound back to Washington, with an overnight stop at Naples. The next morning, General Kelley visited Marines and sailors in the Naval Hospital in Naples, where he presented Purple Hearts to the injured.

During the return trip, the Commandant prepared a hand-written, 23-page outline on a yellow, legal-size tablet, which would be the basis for his official report to the President.

Immediately upon his return to Washington, the Commandant was advised that beginning Monday, 31 October, he would participate as the principal witness in a number of congressional hearings on the Beirut bombing. In preparation, during the weekend of 29-30 October, General Kelley participated in round-the-clock meetings and discussions with Lieutenant General Miller, the FMFLant commander; 2d Marine Division commander, Major General Gray; and his principal staff members to assist him in the preparation of his appearance on the Hill. His primary guidance to his advisors was that regardless of the consequences, "the Marine Corps must tell it like it is."[74]

On Capitol Hill, General Kelley opened his 20-page statement by saying that the attacks on 23 October in Beirut were not against just the Marines and the French, but against the free world.[75]

He then gave his reasons for requesting the Secretary of Defense to establish an independent inquiry into the events leading up to the bombing. The Commandant set the Beirut scene the Marines faced and described their mission (which he said was not "a classic military mission") giving the background for that mission. He avoided discussion of the political or diplomatic reasons for the Marine presence in Lebanon. "It is not the place of a Marine to discuss those imperatives for military employment."[76]

General Kelley spoke about the mission of "presence" and what it meant to the Marine commanders of the MAU. He outlined what each deployed MAU had faced in Beirut, and how the situation was subject to constant change with no corresponding change of mission. The Commandant then discussed the phase that began on 26 September 1983, when a ceasefire had been declared, and when warning of a terrorist threat had been raised again by the intelligence community.

He pointed out that since 1 June 1983 over 100 car bomb possibilities had been developed. All the makes, colors, and license plate numbers of these cars were provided the Marines by intelligence sources and protective measures were taken. General Kelley described what took place on Sunday morning, 23 October, and why he believed that "only extraordinary security could have met the massive and unanticipated threat."[77] Continuing, he said that he believed that Beirut Marines had been pinpointed for destruction by non-Lebanese elements. He then listed the initiatives that were underway or contemplated for increased security of the MAU. He also said that the 24th MAU was in the process of decreasing the vulnerability generally associated with large troop concentrations. Specifically, the steps being taken were to:

> Position Lebanese Armed Forces armored personnel carriers at the BIA terminal and at the traffic circle in front of the airport.

Restrict vehicle access to command posts to emergency and military vehicles only.

Restrict civilian pedestrian access to the command posts to one location.

Have MAU units in an indefinite "Condition 1" (highest level) alert status.

Block and reinforce all entrances to the command posts.

Reinforce all perimeter fence lines adjacent to rifle company positions.

Position an additional .50 caliber machine gun to cover avenues of approach into the MAU command post.

Establish additional guard posts throughout the MAU area, and to request an additional rifle company from Camp Lejeune to provide security during the period of the recovery operations.

Establish mobile reconnaissance patrols with antitank weapons within the BIA perimeter.

General Kelley further noted that Headquarters, 2d Battalion, 6th Marines command elements had arrived at Beirut on 25 October to replace the BLT 1/8 command elements and that definitive action was underway to strengthen the 24th MAU positions and to reduce vulnerability to terrorist attacks by isolating and barricading command and control and support areas.[78]

In summary, General Kelley said:

> Our security measures were not adequate to stop a large heavily laden truck, loaded with 5,000 pounds of high explosive, travelling at a high speed and driven by a suicide driver, which executed the attack in seconds from start to finish. This 'flying truck bomb' was an unprecedented escalation of the previous terrorist threat, both in size of the weapon and method of delivery. I must continue to emphasize, however, that under our current disposition, restrictions, and mission, we will always have vulnerabilities, and that the other side will make every effort to exploit them.[79]

He then added two final comments. The first had to do with a news story which said that Colonel Geraghty had received a warning of the bombing threat two days before the incident. In response to that report, General Kelley quoted Colonel Geraghty, who said:

> Comment made to media was a general statement on car bomb warning. At the weekly intelligence meeting between MNF Intel Officers and the Office of Beirut Security, a listing of suspect car bombs, complete with car descriptions and license plate numbers, is disseminated to the MNF by security officials. These car descriptions are copied and disseminated to our posts. Since our arrival, at least 100 potential car bombs have been identified to the MNF. After the attack on our convoy on 19 October 1983, the car bomb threat was quite obviously real to the USMNF; however, specific information on how car bomb attacks were to be conducted (i.e, kamikaze) or a description of the large truck that conducted the attack on the BLT were never received by the 24th MAU.[80]

The Commandant addressed the charge that, when he was asked in Beirut the previous week whether he thought that security was adequate, he had answered "yes" despite the undeniable fact of the bombing. He explained:

USMC Photo by LCpl Brenda Kusay

MajGen Alfred M. Gray, 2d Marine Division commander, addresses the more than 5,000 people attending the memorial services at Camp Lejeune, North Carolina, on 4 November. Present were President and Mrs. Reagan, Gen Kelley, and other dignitaries.

> Five thousand pounds of high explosives destroyed a four story steel reinforced concrete building. It was a heap of rubble. For over fifty hours, day and night, young Marines clawed at steel and concrete—more to save the injured who were trapped at the time than to recover the dead. The emotional scars were already deep—'Why me?' they asked. 'Why am I alive and my buddies dead?'
>
> Their Commandant was asked, 'was security adequate?' I replied yes—it was adequate to meet what any reasonable and prudent commander should have expected prior to dawn on Sunday, October 23, 1983. And I want you to know in that atmosphere my remarks were directed to weary and frustrated Marines.
>
> Let me phrase what I was saying in a different way: If you were to ask whether the security around the headquarters building was adequate to protect the occupants against a 5-ton Mercedes truck carrying 5,000 pounds of explosives at high speed—my answer would be NO!
>
> And, if you would ask me whether the commander should have known, given the explosion in the Embassy in April, my answer would again be NO! Both instances involved a terrorist bombing from a motor vehicle, but there the

USMC Photo by LCpl Brenda Kusay

Rainsoaked Marines attending the memorial services at Camp Lejeune pay a final tribute to Marines, soldiers, and sailors who died in Beirut and Grenada.

> similarity ends. The delivery system was totally different as was every other aspect of the two incidents.
>
> For these reasons, Mr. Chairman, I urgently requested the inquiry previously mentioned to determine the facts in an atmosphere that is conducive to such an inquiry. Knowing the Secretary of Defense as I do, and the respect I have for Admiral Long, there is no question in my mind that it will be a complete and thorough examination of this awful tragedy. I suggest we all await the board's findings.[81]

Another matter of national concern to which General Kelley addressed his remarks was the manner in which the Marine Corps reported its casualties. He pointed out that in the impact of the destruction of the BLT building and the subsequent tragic loss of life, the casualty reporting procedures for BLT 1/8 was delayed. It was necessary to proceed slowly in reporting for the sake of accurate identification and proper notification to the next of kin of the dead Marines, as well as the wounded. Because of the size of the task facing the Marine Corps:

> . . . and the painfully slow progress in this regard, the decision was made to release the names of those Marines who survived this disaster. We did not do this before for obvious reasons. The process was slow, mainly because of the need for complete accuracy. We didn't want to hurt anyone needlessly. Marines and members of your staffs worked tirelessly to ensure that timely and accurate information was released. The enormity of the situation is still upon us, and no one could feel more remorse than I do over the prolonged suffering caused to many families by unavoidable delays in notifying them of their loved one's status.
>
> The Marine Corps is proud of many things, but nothing more than the way we take care of our own. I want each of you to know that everything humanly possible is being done to facilitate the process.[82]

The Commandant ended his testimony by saying, "The perpetrators and supporters of this challenge to the rights of free men everywhere must be identified and punished. I will have little sleep until this happens."[83] Shortly after the Commandant's Capitol Hill appearance, a commission headed by Admiral Robert L. J. Long, USN (Ret.) was appointed by the Secretary of Defense and began its investigation.

On 4 November, the Commandant accompanied President and Mrs. Reagan and other high government officials to a nationally televised memorial service held outdoors in a heavy downpour of rain at Camp Lejeune. Five days later, General Kelley sent the following message to the 24th MAU, still in Beirut:

> Subj: Outpouring of Concern for Lebanon Marines
>
> 1. Since the tragic events of 23 October there has been an outpouring of concern from people and organizations throughout the world for you. From small towns in middle American to the far corners of the world, I have received hundreds of letters and telegrams from sympathetic and appreciative individuals and organizations expressing their heartfelt concern for Marines and sailors of 24th MAU and your families and their deep appreciation for your sacrfices and your continued dedication to duty.
>
> 2. The tremendous volume of letters and telegrams received precludes presenting extracts from even a fraction of them; however, I have chosen a few salient ones that I felt were worthy of passing on.
>
> A. From the Commandant General Royal Marines, "All Royal Marines grieve with you over your losses in Beirut and your families are very much in our minds."
>
> B. From Northside High School, Memphis, TN, "Northside Cougars care for our Marines in Beirut. . . . We send our love and prayers."
>
> C. From a former Marine in Alabama, "Want you to know that we support you and all your endeavors. We want you to know that if we are not with you in body, we are with you in spirit."
>
> D. From the Swiss Military Attache, "I'm shocked by this terrible act of violence and would like to express to you my condolences."
>
> E. From LtGen Park Hee Jae, Commandant of the Korean Marine Corps, "ROK Marines offer their condolences to those U.S. Marines who sacrified their lives for peace and freedom."
>
> F. From the Mayor of St. Petersburg Beach, FL, "The city government and its employees wish to extend their deepest sympathies for the loss of American lives in Beirut."

G. From the Comanding Officer, 1stBn, Royal Welch Fusiliers, "Our deepest commiserations on your recent losses. . . ."

H. From a young woman in Milwaukee, WI, "May God watch over all of you."

3. Similar messages were received from the German Navy, the Brazilian Marine Corps, NATO, Retired Dutch Marines, and a host of other sources throughout the United States and around the world. It is most heartening to know that so many people outside our Corps care so much for our Marines and sailors and understand and appreciate the difficult and demanding mission that has been given to 24th MAU.

4. As always you and your brave men are in my thoughts and prayers. God bless you and Semper Fidelis![84]

CHAPTER 7
The Investigation

As a result of the Commandant's request for an investigation of the bombing incident by an independent commission, on 7 November 1983, Secretary of Defense Weinberger convened The DOD Commission on Beirut International Airport (BIA) Terrorist Act of 23 October 1983. This move was taken in accordance with the provisions of the Federal Advisory Commission Act (Public Law 92-463). The commission's proceedings were to be governed by Executive Order 12024 and by General Services Administration and Department of Defense regulations.

The charter of the commission stipulated that its advisory function was to be completed within 90 days. Appointed members were Admiral Long; The Honorable Robert J. Murray; Lieutenant General Lawrence F. Snowden, USMC (Ret.); Lieutenant General Eugene F. Tighe, Jr., USAF (Ret.); and Lieutenant General Joseph T. Palastra, Jr., USA.

Also assigned to the commission were both military and civilian assistants to provide advice in various technical areas which would be investigated. Intelligence, planning, operations, special warfare, terrorism, command relations, medical, and international law experts were assigned as full-time staff assistance. Since the substantive information to be collected necessarily included highly classified national security material and because these matters could not reasonably be segregated into separate classified and unclassified categories, all witnesses were interviewed in closed sessions.

The Long Commission visited all major units in the chain of command—from the 24th MAU ashore in Beirut; to Task Force 61 offshore of Lebanon; to Sixth Fleet on board *Puget Sound* at Gaeta, Italy; to CinCUSNavEur in London; to CinCEur in Stuttgart. Commission members also visited Tel Aviv, Rota, Akrotiri, and Wiesbaden. While in Beirut, the commission met with members of the 24th MAU before it left Lebanon, toured Marine positions along the airport perimeter, and inspected the ruins of the BLT building. In Lebanon as well, commission and staff members met with Ambassador Bartholomew and his embassy staff, General Tannous—Commander of the LAF—and the commanders of the French, Italian, and British MNF contingents.

Some of its preliminary findings were time-sensitive. Upon the commission's return from Beirut, it provided the Secretary of Defense with a memorandum outlining the 24th MAU's existing security arrangements. The commission also sent a second memorandum to the Chairman of the Joint Chiefs regarding the Federal Bureau of Investigation's comprehensive report on the nature of the explosive devices used in the embassy and BLT building bombings, with a recommendation that the report be forwarded to the service chiefs as well.

In the belief that a thorough understanding of the circumstances surrounding the bombing of the BLT headquarters required a comprehensive knowledge of a number of separate, but closely related substantive areas, the Commission divided its report into ten parts:

> Part 1: Addresses the development of the mission assigned to the U.S. Multi-National Force, assesses the clarity of the mission, and analyzes the continued validity of the assumptions upon which the mission was based.
>
> Part 2: Addresses the adequacy of the rules of engagement that governed the execution of the mission.
>
> Part 3: Outlines the chain of command that was tasked with the accomplishment of the military mission and assesses its responsiveness to the security requirements of the MAU in the changing threat environment.
>
> Part 4: Examines the threat to the MAU, both before and after the attack, and assesses the adequacy of the intelligence provided to Colonel Geraghty.
>
> Part 5: Analyzes the security measures that were in force prior to the attack.
>
> Part 6: Provides a comprehensive recapitulation of the tragic events of 23 October 1983.
>
> Part 7: Describes the security measures instituted subseqent to the bombing and assesses the adequacy.
>
> Part 8: Reconstructs and evaluates on-scene casualty handling procedures, aeromedical evacuation, and definitive medical care provided to the victims of the attack. Also addresses the circumstances surrounding an Israeli offer of medical assistance and examines the basis for its non-acceptance.
>
> Part 9: Addresses the 23 October 1983 bombing in the context of international terrorism and assesses the readiness of U.S. military forces to cope with the terrorist threat.
>
> Part 10: Lists the commission's major conclusions and recommendations.

The commission's philosophy in preparing the report was outlined:

> . . . the Commission analyzed those factors bearing upon the security of the USMNF in Lebanon in general, and in the security of the BLT Headquarters building in particular. The Commission began with the premise that U.S. par-

Department of Defense photo by Frank Hall

Members of the Long Commission pose with Secretary of Defense Caspar Weinberger before beginning their hearings into the bombing: (left to right), LtGen Joseph T. Palastra, Jr., USA; Adm Robert L. J. Long, USN (Ret); LtGen Lawrence G. Snowden, USMC (Ret); Mr. Weinberger; LtGen Eugene F. Tighe, Jr., USAF (Ret); and Mr. Robert J. Murray.

ticipation in the Multinational Force was designed to support the efforts of the United States and its allies to facilitate the withdrawal of foreign military forces from Lebanon and to assist the Lebanese Government in establishing sovereignty and authority over the Beirut area. The Commission did not question the political decision to insert the Marines into Lebanon and did not address the political necessity of their continued participation in the Multi-National Force following the 23 October 1983 terrorist attack. Although those political judgements are beyond the purview of the Commission's Charter, and not addressed in the report, the fact did not impede the work of the Commission in examining the impact of those policy decisions on the security of the USMNF.

The Commission reviewed the responsiveness of the military chain of command as it pertained to the security requirements of the USMNF. The Commission did not conduct an administrative inspection of any headquarters element during the review process.

The Commission's focus was on the bombing of 23 October 1983 and the security of the USMNF both prior to and subsequent to that catastrophic event. The security of offshore supporting forces was not reviewed in depth by the Commission. The security of other American personnel in Lebanon was not considered, being outside the Commission's Charter.[2]

The report that the commission delivered to the Secretary of Defense was one of the most comprehensive studies prepared on the history and development of the U.S. presence in Lebanon and the rationale for the Marines' mission.

This history will not review in full the lengthy study the Long Commission published on 20 December 1983. Part 10 of the report, "Conclusions and Recommendations," can be found in Appendix E. The main areas investigated by the commission covered the Beirut bombing in general, as well as certain factors affecting the MAUs over which they had no direct control.

The commission concluded that the so-called "presence" mission was not interpreted the same way by all levels of command. These differences, including the responsibilities of the Marines for the security of Beirut International Airport, should have been recognized and corrected within the chain of command. On the expanding role of the MAUs, the commission concluded that high-level decisions regarding Lebanon were characterized by an emphasis on military options and expansion of the roles, despite the fact that the security of the Marines continued to deteriorate as progress toward a diplomatic solution slowed. Decisions affecting the role of the MAUs were taken without clear understanding that the conditions under which the Marines first deployed to Lebanon had changed; that the nature of the American military involvement in Lebanon had changed, and that the expansion of our military involvement in Lebanon greatly increased the risks of the Marines. The commission then recommended a re-examination of alternative means of achieving American objectives in Lebanon, "to include a comprehensive assessment of the military chain of command and a more vigorous and demanding approach to pursuing diplomatic alternatives."[3]

With respect to the rules of engagement, the commission found that a specific set of rules for countering the types of terrorist attacks committed against the embassy and the BLT building had not been provided to nor implemented by the MAU commanders. In addition, the commission said that the Marine mission statement, and the implementation of the May 1983 dual "Blue Card-White Card" rules of engagement, contributed to a mindset which detracted from the Marines' readiness to respond to the type of terrorist attack which occurred on 23 October.

The commission was critical of the chain of command, finding it deficient in several ways:

> 1. An effective command supervision of the MNF security posture was lacking prior to 23 October.
>
> 2. The operational chain of command's failure to correct or amend the Marines' defensive posture constituted tacit approval of the security measures and procedures in force at the BLT headquarters building on 23 October.
>
> 3. Although the USCinCEur operational chain of command was at fault, a series of circumstances beyond the control of these commands influenced their judgement and their actions relating to the MAU's security.

In view of these findings, the commission recommended that the Secretary of Defense ". . . take whatever adminstrative or disciplinary action he deems appropriate, citing the failure of the USCinCEur operational chain of command to monitor and supervise effectively the security measures and procedures employed by the USMNF on 23 October 1983."[4]

Although Colonel Geraghty had received a great volume of intelligence warnings about potential terrorist threats before 23 October, the commission concluded that he was not provided timely intelligence tailored to his specific needs, that would have enabled him to defend against the full spectrum of threats he faced. In addition, the paucity of HUMINT assets—and the fact that the HUMINT he received was neither precise nor tailored to his needs—limited Colonel Geraghty's ability to perceive clearly the severity of the threat he faced. The commission made several recommendations concerning the establishment of an all-source fusion center which would tailor and focus ". . . all-source intelligence support to U.S. military commanders involved in military operations in areas of high threat, conflict, or crisis." The commission also recommended that the Secretary of Defense, together with the Secretary of State and the Director of the Central Intelligence Agency jointly examine current HUMINT activities with a view to improving this type of intelligence support to the American contingent in Lebanon and other U.S. military forces that might operate in areas of potential conflict.

Part 5 of the report's conclusions and recommendations deals with Marine security before the attack and command responsibility for the security of the 24th MAU and BLT 1/8 prior to the attack. The Commission concluded that the security at the MAU compound was neither equal to the increasing level of threat nor sufficient to preclude the catastrophic losses suffered on 23 October. The decision to house approximately one quarter of the BLT in a single structure as a response to incoming hostile fire was found to contribute to the great loss of life. Accordingly, the commission held the BLT commander responsible for placing about 350 members of his command in one building. The MAU commander was held equally responsible for condoning the concentration of troops in the BLT building; for concurring in the BLT commander's modification of prescribed alert procedures; and for emphasizing safety over security, in directing that sentries on posts 4, 5, 6, and 7 carry unloaded weapons. The commission softened these findings by recognizing a series of circumstances beyond the control of both Colonel Geraghty and Lieutenant Colonel Gerlach which influenced their judgement and actions relating to the security of the MAU. Nevertheless, the commission recommended that the Secretary of Defense take adminstrative or disciplinary measures against these two officers.

The commission also discussed post-attack security, noting that actions taken subsequent to 23 October had reduced the vulnerability of the MAU to a similar suicidal attack, but that security measures were still not sufficient to prevent additional casualties from being suffered by the Marines. Although the improved disposition of the Marines may have reflected the best option available, the commission felt that the Department of Defense should prepare and submit to the National Security Council a comprehensive set of alternatives to the status quo in Beirut.

In a discussion of casualty handling, the Commission praised the speed and skill with which the rescue and medical efforts were mounted. It found little to criticize in the aeromedical evacuation of the casualties or their distribution to medical centers, but it pointed out that there were an inadequate number of experienced medical planning officers in the CinCEur chain of command. Another medical aspect had diplomatic implications, for it concerned the rejection of the Israeli offer of medical assistance immediately following the bombing. Commodore France had considered accepting the offer, but determined that the medical capabilities of his command were functioning adequately and that the casualty evacuation plans were being implemented smoothly under actual crisis conditions.

In the report's final section, the commission discussed military response to terrorism. It concluded that the bombing of the BLT building was a terrorist act ". . . sponsored by sovereign States and organized political entities for the purpose of defeating U.S. objectives in Lebanon." It also concluded that international terrorist acts like those which occur in the Middle East constitute a world-wide threat to American and other facilities. Terrorism, the commission concluded, has become an important part of warfare and it is necessary to develop an active national policy to combat it and reduce its effectiveness. The members called upon the Secretary of Defense and the Joint Chiefs of Staff to develop appropriate military responses to terrorism and to work with the National Security Council to develop appropriate political and diplomatic measures.

In conclusion, the commission stated that the Marine force in Lebanon was not trained, organized, staffed, or supported to deal effectively with the terrorist threat in that country. It called upon the Secretary of Defense to ". . . direct the development of doctrine, planning, organization, force structure, education, and training necessary to defend against and counter terrorism."

On 30 December 1983, in response to the Long Commission report, the Secretary of Defense signed a number of memoranda addressed individually to the Secretaries of the Army and the Navy, the Assistant Secretary of Defense for International Security Affairs, the Deputy Under Secretary of Defense for Policy, and the Chairman of the Joint Chiefs of Staff. Each addressee was referred to a specific portion of the report which came within his province and was requested to report to Mr. Weinberger by 9 January 1984 what action he was taking to correct deficiencies or to implement the recommendations made by Admiral Long and his colleagues. The Service secretaries were referred to the appropriate parts of the report in which the Commission recommended that administrative or disciplinary action be taken with regards to individuals, but in his 30 December memorandum, the Secretary of Defense mentioned only ". . . administrative action."

The memorandum to the Chairman of the Joint Chiefs of Staff dealt with several separate topics, and General Vessey furnished copies to each Service chief for information. The Chairman was asked to outline actions he had taken with respect to the commission's recommendations about military responses to terrorism; casualty reporting; the chain of command and effective command supervision of the USMNF security positions; tailored intelligence; rules of engagement; and post-attack security.

Meanwhile, the House Armed Services Committee (HASC) mounted its own investigation. The subject of the HASC Investigation Subcommittee's effort was "Adequacy of U.S. Marine Corps Security in Beirut." The subcommittee was tasked to examine the U.S. policy objective in Lebanon; how the Marine mission contributed to those objectives; whether the risks to the Marines were adequately assessed; and whether adequate precautions were taken to counter them.[5]

On 12 November, the subcommittee delegation arrived in Beirut to conduct two days of hearings, in which they interviewed Commodore France; Colonel Geraghty; embassy security officer Alan O. Bigler; Commander Richard Balzer and Lieutenant Frazer Henderson, both Navy doctors; Lance Corporal Berthiaume; and Navy Hospital Corpsmen Michael Arau and Donald Davidson.

Earlier that month, the full committee held two days of hearings in Washington, during which time the following testified: General Kelley; General Mead; Captain Lewis Mantel, a Navy doctor; Congressman G. V. "Sonny" Montgomery; Rear Admiral Jonathan T. Howe, Director of the Bureau of Politico-Military Affairs at the State Department; Ambassador Dillon; and Gordon E. Harvey, Deputy Director of the State Department's Office of Security. In early December, the Subcommittee also heard from Ambassadors Dillon and Habib; four witnesses from the National Security Agency; John W. Hicks, Chief of the Scientific Analysis Section, Federal Bureau of Investigation; Legislative Counsel Jack Perkins, Office of Legislative Affairs, Department of Justice; General Mead again; Corporal Joseph Martucci; Lance Corporals Burnham Matthews and Robert Calhoun; and First Lieutenant Gregory P. Balzer.[6]

On 14 and 15 December, the Subcommittee heard testimony from Aviation Electronics Technician Talmadge E. Lea; First Lieutenant Glenn L. Wagner; Commodore France; Colonel Geraghty; Petty Officer Kenneth W. Densmore; General John W. Vessey, Jr.;

and General Bernard Rogers, CinCEur. The Subcommittee prepared its report from sworn testimony at these hearings and it was published in two sections—the main report itself with diagrams, maps, photographs, and selected portions of testimony, and a second section that contained a summary of findings and conclusions. Both sections were published and made available on 19 December, a day ahead of the Long Commission report.

The Subcommittee found that inadequate security measures had been taken to protect the MAU from the full spectrum of threats. In addition, Colonel Geraghty was found to have made ". . . serious errors in judgement in failing to provide better protection for his troops within the command authority available to him."[7] Commodore France was adjudged to be equally culpable. The summary also stated that higher command elements failed to exercise sufficient oversight of the MAU. The House members were particularly concerned that the higher level commanders did not reevaluate the security posture of the MAU when it appeared to become increasingly vulnerable in the weeks before the bombing, and that the high level visitors to Beirut, although they were given familiarization briefings, did not seem to be sensitive to the increased security needs of the MAU. The Subcommittee also called into account the role of the ". . . higher level policymaking authority that adopted and continued a policy that placed military units in a deployment where protection was inevitably inadequate."[8] The Subcommittee's summary continued:

> Both the Marine ground commanders who testified, consistent with the view of the Marine Corps leadership, interpreted the political/diplomatic nature of the mission to place high priority on visibility and emphasized to the extent of allowing greater than necessary security risks. The subcommittee was particularly distressed to find that the security of the MAU was less than that provided at the interim U.S. Embassy in Beirut.[9]

The Marines in the MAU were praised for their skill, courage, and fortitude, and were considered to be functioning well in a role that was less military than diplomatic. Continuing, the summary dealt with intelligence matters and found that the MAU did not receive adequate intelligence about terrorism, and that the MAU erred in failing to consider truck bombs as significant threats.

In conclusion, the subcommittee in strong terms urged the Administration to review its policy in Lebanon, ". . . from the standpoint of how the Marine mission fits into the policy to determine if continued deployment of the Marine unit, as part of the Multinational Force (MNF) of French, British, and American units is justified."[10]

Finally, the congressmen stated: "The solution to Lebanon's problem will only be found at the bargaining table. We must not in any way encourage the perception that a solution can be found on the battlefield with the participation of U.S. armed forces."[11]

CHAPTER 8

Beirut V Goes Home

On 26 October, as Vice President Bush visited Beirut, Company B positions took 15 mortar rounds over a two-hour period. They returned the fire with 21 rounds of high-explosive 81mm ammunition. Before the month was over, the MAU would suffer three more wounded, none of whom needed to be evacuated.[1]

As attempts to recover bodies and clean up the site continued, the MAU worked hard to prevent a recurrence of the suicide attack. Earthen- and concrete-filled barricades were placed in all open areas to forestall high speed entry by attackers. Colonel Geraghty established a fortified perimeter within the Beirut airport area. He ringed the MAU command post with an anti-vehicle ditch and an anti-vehicle berm, along with the following:

1. The airport road was reduced to two from four lanes.
2. Access into the perimeter was restricted to Multi-National Force and U.S. Embassy vehicles only.
3. The number of entrances to the MAU command post was reduced to three, all of which were covered by .50 caliber machine guns and blocked either with a five-ton truck or heavy steel gates made of railroad tracks.
4. M-60 machine guns, loaded with 7.62mm armor-piercing ammunition, covered all roads and open areas leading into or in the proximity of the airport area.
5. With the exception of those Lebanese who worked at the airport power plant, all civilian personnel were excluded from the compound and all Lebanese Armed Forces troops were relocated outside the fenceline.
6. The number of interior guard posts was increased and all posts were armed with LAAWs (light antitank assault weapons).
7. All but 10 security guards were removed from Green Beach.
8. The Corniche in front of the Durrafourd Building and the British Embassy was blocked off and the position reinforced by an armored assault vehicle.[2]

The Marines remained alert to the possibility of kamikaze-like tactics by fanatics wearing stolen uniforms and driving captured MNF and LAF vehicles. Fighting continued in the suburbs of Beirut during the last days of October, while "The myriad of intelligence reports involving planned bombings of the MNF and diplomatic locations coupled with rumored U.S. retaliation in the southern suburbs only increased the already high tension in Beirut."[3]

On the 31st, an amtrac was convoyed to the Marine guard positions at the Embassy to provide increased security there. Meanwhile, the newly arrived Company E, BLT 2/6, was inserted into positions on the northeastern portion of the perimeter, relieving Company C, which now moved into the location formerly held by the 155mm howitzer battery. The artillery was moved to new emplacements in the southwestern part of the airport because it had come under heavy fire in the north and was unable to guarantee immediate fire support when called upon to provide it.

By the end of the month, HMM-162 had accumulated 7,435.4 accident-free hours of deployed flight time, exceeding the record of any squadron previously deployed with the Sixth Fleet. A combination of factors, such as medevac, VIP, cargo, mail, and diplomatic flights resulted in this record, which was coupled with a high percentage of aircraft availability, 90 percent, attesting to the round-the-clock efforts of the maintenance crew.

A sorely tried Colonel Geraghty commented at this time:

> While the cutting edge of the MAU took many nicks this week, it proved to be made of well-tempered steel. Those that have tried to dull the blade have found that it can't be done from a distance and they have had to move close aboard. . . . The support from the MARG during the bombing was without equal. They placed their ships in harm's way and were the first to respond. Many Marines owe their lives to the sailors of TF 61.[4]

An FBI study later revealed that the explosion which collapsed the BLT building had been caused by explosive material wrapped around tubes of propane or another type of highly volatile gas. This boosted the explosive force of the bomb to the equivalent of more than 12,000 pounds of TNT. In addition to collapsing the building, this was enough to make a crater measuring 39 feet by 29 1/2 feet and 8 feet deep. In doing this, the explosion destroyed a seven-inch-thick concrete floor, which was reinforced by steel rods, each 1 3/4 inches in diameter. The FBI also concluded that, even if the truck had not reached the lobby, and had exploded instead in the roadway at a distance of 330 feet from the building, nearly the same amount of damage and a significant number of casualties still would have resulted.

A large congressional delegation arrived on 29 Oc-

USMC Photo by SSgt Robert E. Kline

Flanked by American diplomatic personnel as well as representatives from other Multi-National Force units and the Lebanese Armed Forces, Col Timothy J. Geraghty presides over a memorial service in front of the MAU headquarters building in early November.

A camouflage utility cap and artificial flowers poignantly mark what is left of the destroyed Marine headquarters days after the site had been cleared. Not much else remains.

Photo courtesy of Francoise de Mulder

tober and was briefed and given a tour of the MAU positions. Other high-level visitors at the end of October and the beginning of November were General Richard L. Lawson, Deputy Commander, EUCOM; Admiral Richard N. Small, CinCUSNavEur; and Lieutenant General Howard F. Stone, Chief of Staff of EUCOM. On 4 November, the MAU held an emotion-filled memorial service for the men killed in the bombing. Present at the services were Ambassador Bartholomew and Deputy Chief of Mission Pugh and their wives, together with representatives of the LAF and MNF units.[5]

Both Companies A and E came under fire on the night of 5 November, as the advance party of the 22d MAU arrived for the turnover. Beginning at 1650, 7 November, all units on the perimeter came under intense fire, which continued for six hours and was ended only by a heavy downpour of rain and hail.

Because of the difficulty in supporting Company A at the Lebanese University, and the threat these Marines faced by being isolated from the rest of the MAU, Colonel Geraghty decided to abandon the position and pull the company back into his lines. At 0321 on 8 November, a convoy arrived at the university to withdraw Company A. Mechanical difficulties with one of the vehicles delayed the return trip and the convoy did not return to the airport until three and a half hours later. Company A then embarked on board the *Harlan County*, where it remained until her return to the States. This move essentially left the British contingent isolated across from the university on the Old Sidon Road.

The Marine Corps Birthday was celebrated in Beirut on 10 November with a cake-cutting ceremony held at the MAU headquarters. Of this event, Colonel Geraghty wrote, "Our birthday celebration was low key, but traditional, and from our watch, 24th MAU added another page to the history of Marines in Lebanon."[6] On the 10th, Colonel Geraghty received the following message:

> Before long you will turn over your responsibilities to 22d MAU. All Americans are deeply in your debt. Even as we grieve for your sacrifice, we take pride in your excellence as Marines on this, the 208th Birthday of the Corps. Please know we are thinking of you and look forward to welcoming you home—Our Marines. 'Semper Fidelis,' Ronald Reagan.[7]

With a week left until its relief by the 22d MAU on 19 November the 24th MAU's Marines remained on alert, receiving some intermittent fire and returning it when warranted, but ready to leave Lebanon. Morale was very high, but the shock of the bombing and the resultant loss of life began to tell within a few days after the 23d. For many of the younger Marines, who had never before faced death or the dying so closely, it was a traumatic experience. Chaplain Pucciarelli counseled many Marines who had lost good friends, and even relatives, in the bombing. Some of the older and more mature Marines spoke to the younger ones, trying to get them to talk about their feelings to provide an emotional outlet. With respect to this period and the bombing, the MAU chaplain later reflected:

> You know, we can read about Pearl Harbor, you see pictures, but being there [in Beirut] again, the sights and the sounds and the smells and all the senses would be more of a sobering event than reading this in a history book. So a lot of these young lads were overcome by the scene, and of course, it took its toll that particular day. But I think, working at the site for four days or more, helped a lot. I think it was kind of a [catharsis].[8]

Elements of Lieutenant Colonel Ray L. Smith's BLT 2/8, fresh and eager after a successful operation in Grenada, began landing at Beirut International Airport on 17 November. At that time, BLT 1/8 began to backload on Phibron 8 shipping. By 2330 the next day, all members of the 24th AMU were re-embarked and ready to leave for home. Brigadier General Jim R. Joy,* the 22d MAU commander, relieved Colonel Geraghty as commander of the U.S. contingent of the Multi-National Force, Beirut at 1000 on 19 November. Two hours later, the *Iwo Jima* and the *Portland* followed in the wakes of the *El Paso, Harlan County*, and *Austin*, which had left for Rota, Spain the day before. In his last situation report from Lebanon, Colonel Geraghty stated, "24th MAU stands relieved as LF6F 2-83 [Landing Force, Sixth Fleet] and US Contingent to the Multi-National Force, Beirut, Lebanon. Proceeding on duties as assigned. Able to respond to any combat mission. Able to respond as Marines."[9]

Phibron 8 arrived in Rota on 24 and 25 November to a warm reception that was totally unexpected. The naval station commander extended post exchange hours to accommodate MAU/MARG personnel and to ensure taht they had an opportunity to relax and shop.

*In view of the need to provide additional supervisory assistance and coordination of the activities ashore in Beirut, the Commandant authorized the appointment of the Assistant Commander, 2d Marine Division, Brigadier General Joy, as commanding general of the 22d MAU on 3 November. When Phibron 4 shipping carrying the 22d MAU from Grenada to Beirut arrived off Lebanon on 17 November, General Joy went on board *Guam* and formally assumed command of the 22d, relieving Colonel Faulkner, who then became chief of staff. Prior to his promotion, General Joy had served as Fleet Marine Officer, Sixth Fleet and was totally familiar with the Lebanon situation.

USMC Photo by SSgt Robert E. Kline

24th Marine Amphibious Unit Marines wash down their vehicle at Beirut International Airport before it went on board Phibron 8 shipping, headed for Rota and the States.

Bus transportation was provided to and from the ships on a regular schedule for liberty parties. The wives' clubs at the base organized "welcome home" parties where ". . . tons of fine food of the most enjoyable sort, free beer/soda, a band and singers, and massive amount of good will and friendliness were dispensed by these charming hostesses in a gracious and warm manner. It was a reception which cannot be topped for the amount of care shown. The overflowing of concern was unexpected and deeply appreciated."[10]

On 29 November, while in the Atlantic heading for Morehead City, the 24th MAU received the following message from the Commandant:

> Subj: USMNF
>
> 1. Courage, sacrifice and heroism characterized the Lebanon tour of 24 MAU. Under the most trying and difficult conditions each unit's performance shines as a witness to the world that Americans stood firm in the defense of peace and freedom.
>
> 2. The exemplary bravery of the MAU's Marines, sailors, and soldiers has been indelibly written on the pages of American history. No one—standing or fallen—served in vain. Every man's devotion to duty will continue to be an inspiration to all who desire to live as free men.

BGen Jim R. Joy (right) relieves Col James P. Faulkner as 22d Marine Amphibious Unit commander on board the Guam *(LPH 9) off Beirut on 17 November 1983.*

Photograph by author

3. On behalf of a grateful nation, I thank God for men like you in the service of this country.

General Kelley sends.[11]

The 24th Marine Amphibious Unit arrived at Morehead City to a warm welcome from the commanding generals of the 2d Marine Division, 2d Marine Aircraft Wing, and 2d Force Service Support Group, plus families, bands, and national media attention.

Two days later at Camp Lejeune, Colonel Geraghty's 24th MAU passed in review before General Kelley, who welcomed the Marines and sailors home with the following remarks:

When I met the first flight of your fallen comrades as they arrived at Dover, Delaware, after the mass murder of 23 October, I asked the question—Lord, where do we get such men? As you stand here today I ask the same question. Where do we get such men of courage—such men of dedication—such men of patriotism—such men of pride? The simple answer is that we get them from every clime and place—from every race—from every creed—and from every color. But each of you has one thing in common—you are a Marine or that special brand of Navy man who serves alongside Marines.

Two days ago an entire nation opened its heart in grateful recognition of your safe return.

You gallant Marines and sailors of the 24th have earned your rightful place in the glorious history of our Corps. You can stand tall and proud in the knowledge that you have selflessly given of yourselves in the service of your country, your Corps, and of free men everywhere.

In the joy and emotion of your safe return, let none of us forget those brave Marines and sailors who made the supreme sacrifice—or forget the wife who will never again see her husband—the child who will never see its father—or the parents who will never see their son. They, too, have made the supreme sacrifice!

By the authority given to me this day by the Secretary of the Navy, I hereby recognize your significant contributions, under conditions of great adversity, by authorizing each of you to wear the Combat Action Ribbon.

You and your precious families—those loved ones who have participated in a lonely and anxious vigil these past months—have my deepest and sincerest respect and admiration. God bless you!

CHAPTER 9

Beirut VI—End of the USMNF 20 November 1983-26 February 1984

On 2 August 1983, prior to its return to Lebanon, the 22d Marine Amphibious Unit, now commanded by Colonel James P. "Pat" Faulkner, once again came under the operational control of Fleet Marine Force, Atlantic. At this time, the MAU consisted of BLT 2/8 (LtCol Ray L. Smith), HMM-261 (LtCol Granville R. Amos), and MSSG 22 (Maj Albert E. Shively). All the MAU's Marines, and their Navy colleagues, conducted the usual pre-deployment training and exercises, and boarded Commodore (Captain, USN) Carl R. Erie's Phibron 4 shipping at Morehead City on 17-18 October 1983 for the trip to Beirut. The squadron's ships consisted of the *Guam* (LPH 9), the flagship; the *Trenton* (LPD 4); the *Fort Snelling* (LSD 30); the *Manitowoc* (LST 1180); and the *Barnstable County* (LST 1197). On 18 October, they stood out of the North Carolina port city for Beirut.

For Beirut VI, the 22d MAU's third deployment to Lebanon, HMM-261 and MSSG 22 were organized like their predecessors, but BLT 2/8 had been reorganized to conform to a new infantry battalion table of organization (1083C). This new T/O reduced the Marine infantry battalion by 10 percent, to a strength of 43 officers and 779 enlisted Marines. Despite this reduction, the new battalions were given greater fire power with an increase of 24 grenade launchers (bringing the total to 134), 8 additional Dragon antitank weapons (for a total of 32), and the introduction of 8 M-2 .50 caliber machine guns. At a future date, each of the new infantry battalions would be issued other new weapons—the SMAW (Shoulder Launched Multipurpose Weapon), and the Mk 19 40mm machine gun.

To transport the additional heavy weapons and to give the reorganized battalions greater mobility, they were issued 26 additional jeeps, essentially doubling their previous allowance. In a battalion landing team configuration, the infantry battalions would also gain 24 more jeeps from their attached units.

The reduced strength of the battalions was reflected primarily in the reorganization of the rifle units. A rifle platoon now consisted of 36 Marines—including the platoon leader, platoon sergeant, and platoon guide and 11-man squads of two 5-man fire teams each—instead of the 13-man squads of three four-man fire teams each.

Some of the MAU's Marines had been on an earlier deployment to Lebanon, for more than 40 percent of the BLT had been in the unit two years or more. All of Lieutenant Colonel Smith's squad leaders and more than one third of his fire team leaders had completed the 2d Marine Division's Squad Leader's Course. All the BLT's rifle platoon commanders had been through the Infantry Officer's Course at the Marine Corps Development and Education Command at Quantico following their graduation from The Basic School.[1]

About midnight of 20-21 October, as Phibron 4 shipping passed north of Bermuda en route to the Mediterranean, CinCLant ordered Commodore Erie to turn south to a holding position about 500 miles northeast of Grenada. Because the Phibron's ships' radios had been monitoring the news stories as well as receiving updated classified intelligence reports about the civil upheaval in Grenada, both Navy and Marine Corps officers presumed that they might be directed to conduct a non-combatant evacuation of American and foreign nationals from the troubled island. The Amphibious Task Force had trained for this type of operation and began planning to carry out such an evacuation shortly.

In his message to Commodore Erie, the Commander in Chief of the Atlantic Fleet also instructed the Phibron commander to remain in his holding position until midnight of 23-24 October. Then, if no further instructions had been received, he was to continue on his way to Beirut. At the same time, the Phibron assumed an EmCom (emission control) condition, in which radio and radar silence was instituted. Messages could be received, but not sent, as all electronic and sonic emissions closed down. As Phibron 4 essentially became a ghost squadron, the inability to talk to higher echelons was to cause some problems as planning for the Grenada operation unfolded.

At this point, Marine and Navy planning was primarily concerned with the evacuation of civilians from a hostile or "non-permissive" environment. At 2200 on 22 October, Commodore Erie was ordered to head his ships towards Grenada. A second message then gave order of battle information about Grenadian forces. No further directives were issued to the Phibron at this point. Admiral Joseph Metcalfe III, Second Fleet

Photograph by the author

En route to Beirut from Grenada on board Trenton *(LPD 4) (left to right) Maj Joseph J. Streitz, ExO, BLT 2/8; LtCol Ray L. Smith, CO, BLT 2/8; and Maj Albert E. Shively, CO, MAU Service Support Group 22.*

commander and joint task force commander for the operation, radioed Commodore Erie that the Army had been ordered to conduct an airborne assault on Grenada.[2] Shortly after this, the Marines were brought into the picture and given their operation orders for the landing on Grenada.

By 1 November the 22d MAU had successfully completed its role in Operation Urgent Fury, the code name for the landing on Grenada. On 2 November, Lieutenant Colonel Ronald R. Rice, 22d MAU executive officer, led an advance party ashore on Grenada, where it boarded a plane for the United States and then on to Beirut. The next day, Colonel Faulkner and his operations officer, Major (later lieutenant colonel) Earnest A. Van Huss, were flown ashore to Grenada, where they briefed Senator John G. Tower of Texas on Operation Urgent Fury. Shortly after this, the two flew to Norfolk to brief Lieutenant General Miller at FMFLant headquarters.

At 1740 on the 2d, Phibron 4 ships steamed past St. George's harbor with battle flags flying. The ships then turned and headed north for Barbados, where HMM-261 helicopters flew supplies from the beach to the carrier *Independence*. When this task was completed and all helicopters had landed back on the *Guam*, the Amphibious Ready Group set a course for Beirut.

On 3 November, the 22d MAU received a message stating that when the MAU arrived off Beirut, its structure would be modified.[3] Essentially, Brigadier General Jim R. Joy, the Assistant Division Commander of the 2d Marine Division would relieve Colonel Faulkner as commander of the MAU, whereupon the former commander would become MAU chief of staff. General Joy was to bring a small staff group to Lebanon to expand the MAU staff. The rationale behind this high-level decision was the need to provide additional supervisory assistance and coordination of activities ashore in Beirut. In considering the terrorist bombing of the BLT building and the subsequent recovery measures, as well as the need to coordinate the overall efforts of the other Multi-National Force units and to supervise the relief of the 24th MAU by the 22d, it was deemed necessary to assign a Marine general officer as MAU commander. Additionally, this would make him co-equal in rank to the French and Italian MNF commanders.

General Joy later gave an additional reason for the change. In response to the heavy fighting in late August and early September, the 31st MAU was sent to Beirut from Kenya to serve ashore as theater reinforcement, if needed. At this time, General Miller had directed General Joy at Camp Lejeune to put together a "mini-MAB" [Marine Amphibious Brigade] headquarters, ready to fly out to Beirut should the U.S. Multi-National Force be increased to MAB size.[4]

General Joy then organized what he called a "suitcase staff," consisting of no more than 10 people, which was packed and ready to fly to Beirut when ordered.*

General Joy's small staff was briefed in Norfolk at FMFLant headquarters, and in Washington by Headquarters, U.S. Marine Corps staff sections, by the State Department, and by the Defense Intelligence Agen-

*When General Joy took command of the 22d MAU on 17 November 1983 at 1100, the MAU's staff sections—S-1, S-2, S-3, and S-4—became G-sections. The 22d's S-1, First Lieutenant Kenneth R. Bergman, remained as G-1, and his section was augmented by a warrant officer, whose initial assignment was to work with BLT 1/8 to get its personnel records organized before the unit returned to the United States. Lieutenant Colonel Forrest L. Lucy became G-2, with the former S-2, Captain Paul M. Jungel, becoming his assistant. Similarly, the MAU S-3, Lieutenant Colonel Earnest A. Van Huss became assistant G-3 to Lieutenant Colonel Edmund J. Connelly, Jr., and Lieutenant Colonel Charles S. Rinehart became G-4 with Major Albert J. Martin his assistant. Lieutenant Colonel William H. Schopfel III relieved First Lieutenant Billy D. Martin as the Fire Support Coordinator, and was in turn relieved by Major John R. Todd for seven days, 13-19 February 1984. The only unit commander replaced was Major Albert E. Shively, head of MSSG 22, who became executive officer to Lieutenant Colonel Douglas M. Davidson. When Colonel Faulkner became General Joy's chief of staff, the MAU's former executive officer, Lieutenant Colonel Ronald R. Rice, became MAU liaison officer to the Lebanese Ministry of Defense. On 20 February 1984, Colonel Faulkner again took command of the MAU, and General Joy became Commanding General, Joint Task Force, Lebanon. The senior staff officers who came to Beirut with General Joy became the JTF staff, whereupon their former assistants once again became the MAU's senior staff. 22d MAU Post Deployment Rpt for Landing Force Sixth Fleet 1-84, dtd 11May84, p. 3, hereafter 22d MAU Post Deployment Rpt.

cy. Back at Camp Lejeune, in addition to carrying out its regular assignments, the staff met often to work on contingency plans and to keep current on what was going on in Beirut. The staff remained on alert until early October. When the 31st MAU left Beirut on 13 October to return to the Western Pacific area, the concept of sending General Joy and his staff to Beirut became moot. After the bombing of the BLT headquarters building, however, General Joy was instructed to leave for Beirut as soon as possible and assume command of the 22d MAU before it landed to relieve the 24th.[5]

One day out of Rota, Spain, 10 November, all embarked Marines and their Navy hosts celebrated the 208th Birthday of the Corps in traditional manner, with the reading of Major General Commandant John A. Lejeune's birthday message. Another tradition observed was the cutting of the birthday cake, with the first piece handed to the oldest Marine present, and the second piece to the youngest. Lieutenant Colonel Ray L. Smith, the BLT commander, on the *Trenton*, was to lead the second advance party into Beirut, flying from the *Guam* to Rota on the 11th. As he had to leave the *Trenton* on the 9th,* he held his battalion's birthday ceremonies that morning.

Colonel Faulkner flew on board the *Guam* from Rota on 11 November, and resumed command of the MAU. He then briefed his key staff and command personnel about the new Marine command arrangement that would go into force when they arrived at Beirut and the relief of the 24th MAU.

The Amphibious Ready Group arrived at Beirut on 17 November. General Joy and his staff boarded the *Guam*, where he relieved Colonel Faulkner as 22d MAU commander at approximately 1100.

In early November, before he took over command of the MAU, General Joy was in Beirut to survey the situation. CinCEur sent him a message on 9 November, directing a number of actions to enhance the security of the U.S. Multi-National Forces ashore in Lebanon. Among these was a requirement to reduce the size of the BLT and MAU headquarters ashore to an essential few, with the "non-essential" Marines relocated on board Phibron shipping. Following this, the rifle company at the northern end of the airport would be moved to other positions to provide the MAU with integrated and coordinated security. The company's former positions were to be occupied by LAF troops. Company E, 2d Battalion, 6th Marines, which had

*When the *Guam* headed towards Rota, the rest of Phibron shipping steamed directly into the Mediterreanean, where the *Guam* would join up later.

Photograph by the author

22d MAU commander Col James P. Faulkner (left) and MAU Chaplain Kevin L. Anderson look over a few of the thousands of letters sent to the 22d MAU following its successful operation on Grenada.

reinforced the 24th MAU after the bombing, returned to Camp Lejeune by 19 November. In addition, General Joy was to spread out the concentrated billeting of Marines providing security for the U.S./British Embassy. Further, he was directed to return to shipboard all but the forward (or Alpha) command groups of the BLT and MAU until protected command posts with overhead cover could be constructed for them. Finally, except for a minimum of essential units to provide support ashore, the MSSG was to operate afloat.[6]

Meanwhile, the turnover with 24th MAU went well and was completed 12 hours ahead of schedule on 19 November. General Joy then threw the MAU's entire efforts into improving the safety and security of all troops ashore by constructing additional bunkers, improving existing positions, ensuring dispersion of units, and "fine-tuning the command and control capability of the MAU Hq."[7] The fact that the turnover had gone so smoothly, in perfect weather, and without harassing fire from unfriendly elements, enabled the MAU to push ahead with its barrier and obstacle plan and to begin building a new MAU command post on 19 November.

A Seabee site survey team had been at the airport for two days, 17-19 November, to review the Marine positions and determine how they could be improved and made safer. Meanwhile, the MAU headquarters had been moved to the airport maintenance building just east of its previous site. The new BLT command post was now on a piece of land between the coastal highway and the southern end of the airport's north-south runway. Located on the same stretch of land, but closer to the crossing of the north-south and

U.S. Navy photo

The battleship New Jersey *(BB 62) fires her 16-inch guns off the coast of Beirut in support of Marines ashore when hostile rocket and artillery rounds threatened Americans.*

northeast-southwest runways, were the artillery battery emplacements. Two rifle companies (F and G) were dispersed on the eastern side of the northeast-southwest runway within several hundred yards of LAF and Shiite positions near Khaldah, where the Marines were still subject to frequent sniper fire.

The Seabee report noted that the MAU was attempting to build protective, semi-covered bunkers without enough material, equipment, skilled labor, and experience in constructing such structures. The reporting Seabee officer concluded that these MAU-built bunkers offered little more than minimal protection from shell fragments.[8]

According to General Joy's plan, the MAU command post was to be built near the new BLT command post area. By 19 November, preparation of the site was underway. The possibility of heavy rains in December and the immediate requirement of the MAU to dig in influenced the Seabee survey team leader to recommend that 40 Seabees from the 1st Naval Mobile Construction Battalion Detachment, Rota, be sent to Beirut to assist the Marines in their barrier and construction efforts. The recommendation was approved, the Seabees arrived in Lebanon on 24 November for a 30-day assignment, and immediately began work on the new MAU positions. Initially, the MAU recognized the need to protect its combat operations center, intelligence section, fire support coordination center, and the like. At the same time, General Joy pointed out an equally important requirement for an obstacle/barrier protection system for rifle company positions. The Seabee team recommended that sea-land

Beginning in December 1983, sea-land shipping containers are dug in for use as secure command, control, and communications bunkers at Beirut International Airport.

vans be reinforced and dug in as protected bunkers, a solution which seemed eminently workable. Earlier, the MAU had contracted locally for heavy equipment and construction materials to build/reinforce Marine bunker complexes. At the same time, General Tannous provided the MAU with 40 of the large sea-land shipping containers which the Seabees began reinforcing and burying for MAU, MSSG, battery, and company command and control facilities.

General Joy also requested that upon completion of the construction phase, additional containers be procured and buried for use as protected personnel bunkers. As these construction efforts went on, combat engineers assigned to the BLT assisted the rifle companies in improving and rebuilding their fighting positions. The Seabees were tasked with building "dive-in" bunkers, strong backing for tents, and construction of earth berms between fighting and living positions. The MAU commander wanted to reduce the number of Marines living in buildings in the old MAU/MSSG area, and he predicted in his 19 November report to CinCEur that, when the new MAU command post was completed—within 10-14 days—the number of personnel ashore would be reduced considerably. General Joy concluded his situation report by saying, "We are mindful of requirement for keeping minimum essential people ashore and are reviewing each functional area in the MAU/MSSG on a line by line basis."[9]

From his vantage point in the turret of an LVTP-7 assault amphibious vehicle in front of the British embassy, a 24th MAU Marine keeps watch for potential attacks against the U.S.-British diplomatic center.

Photo courtesy of Claude Salhani

Less than a week after 23 November, General Joy again reported that he, his staff, and his commanders had dedicated their efforts to continuing the "presence" mission while doing their utmost to prevent a recurrence of the bombing and other terrorist actions. At the same time, he recognized that the terrorists might resort to such other tactics as mining the MAU area, and ambushing, kidnapping, or assassinating Marines. The MAU commander further reported that he had identified the Durrafourd Building, the U.S./UK Embassy, and the MAU/MSSG areas as the most likely terrorist targets, and that he had taken the steps he mentioned earlier to protect the Marines against terrorist attacks. To refuse entry into the MAU positions by sappers, infiltrators, and kidnappers, General Joy replaced the fixed positions along the perimeter with aggressive patrolling at irregular intervals. He backed this so-called "forward security line" with section- and platoon-manned strongpoints with mutually supporting crew-served weapons. He also placed tactical and protective wire around the strongpoints and planned to install floodlights at these positions.

General Joy also reduced access to the roads leading into the Marine perimeter with what amounted to a three-tiered system. The innermost tier was armed with direct fire weapons, such as Dragon, LAAW, and .50 caliber machine guns, manned and fully ready 24 hours a day. Each Marine position was issued special rules of engagement based on specific triggering situations that were most likely to occur. The MAU commander had also recognized the potential threat of suicide air attacks and had considered the use of Redeye and Stinger missiles in an air defense role, but because of the danger they might pose to commercial flights in and out of Beirut International Airport, those weapons were not initially used. And so air defense was assigned to .50 caliber and M-60 machine guns.

Finally, General Joy reported that he was fully employing the counterintelligence augmentation he had been given. This consisted of the 2d Counterintelligence Team, a composite team with personnel drawn from the 2d and 4th Counterintelligence Teams (FMFLant), and the 8th Counterintelligence Team (2d Marine Aircraft Wing), augmenting the counterintelligence detachment that originally deployed with the 22d MAU. The composite team operated with a headquarters element and four subteams, each of which was assigned a specific functional area. One subteam

A wrecked automobile is placed with other obstacles in the defense perimeter surrounding the 22d MAU.

was assigned to counterterrorism activities and physical security of the MAU area of operations, while a second was tasked to collect information about threats to the U.S. Multi-National Force. A third subteam was assigned the mission of analysis and reporting, while the fourth was held in reserve with a secondary mission of reinforcing the physical security subteam.[10]

As a matter of Marine Corps doctrine, counterintelligence teams are not normally assigned to or deployed with units below Marine Amphibious Brigade level, and the 22d MAU became the first unit of its kind to be supported by a counterintelligence team that was fully manned and equipped. Lebanon also marked the first time since Vietnam that a counterintelligence team had deployed in support of a landing force commander.[11]

While all this activity was going on ashore, Colonel Faulkner, as chief of staff of the MAU, spent his nights on board the *Guam*, coordinating with the Phibron staff as well as coordinating MAU staff functions afloat. His days ashore were spent at the MAU forward command post at the airport, coordinating MAU staff functions there. This permitted General Joy to devote more time to improving the MAU's defensive positions, "enhancing boring conditions, handling visiting VIPs, and coordinating with other MNF and GOL [Government of Lebanon] agencies."[12]

On 25 November, General Joy reported that two 9-foot berms had been prepared to the north of the MSSG command post and that a tank ditch was being dug in between the berms. At the same time, two 9-foot berms were being built outside the western and southern fence lines encircling the MAU area, after which a tank ditch would be dug inside the fences. Protective wire was strung and the berms were covered by M-60 and .50 caliber machine guns, Dragons, and LAAWs. No Lebanese vehicles were permitted inside the area and all other autos were stopped and inspected before they were given entry. The old gates and weak portions of the fenceline were blocked with wrecked buses and automobiles.

By the 25th, the BLT command post had moved to its new site. On the same day, the Seabee contingent began preparing the MAU headquarters' new bunkers. Earthen berms were thrown up around the BLT and MAU command post sites, and bulldozers were working at the rifle company and artillery battery positions, building berms and clearing fields of fire.

Concurrently with these engineering activities, the rifle company and platoon positions were being reorganized to become mutually supporting. The existing bunkers were used as "passive type" observation and listening posts, while engineer-designed prefabricated fighting positions were placed in the rear of these posts. Once this "frontline" work was completed, the Seabees were to prepare bunkered living positions and sandbagged strong-back tents in the MAU/MSSG and BLT command post areas and at each company and battery position.

In the midst of all this, the MAU remained on alert in order to be immediately responsive to the multifaceted threat it faced. On the perimeter, the Marines were awakened each day for an early morning stand-to and General Joy set Alert Condition 1 in the predawn hours (0445-0700). Fortunately, there had been little or no sniping or incoming artillery and rocket fire during the 22d MAU's first days back in Lebanon.[13]

General Joy maintained close relationships with the other Multi-National Force Units in Lebanon. He proposed setting up a MNF coordinating officer at the Lebanese Ministry of Defense for the then-existing MNF Liaison Office at the Presidential Palace was not working effectively and was not responsive. There was no early decision for or against the proposal, however.

During this period, the MAU's composite helicopter squadron was kept busy with passenger, mail, and freight flights to and from Beirut airport or to Larnaca from the flight deck of the *Guam*. In addition, the helicopters flew VIP shuttle and diplomatic flights, some of which went to Tel Aviv.

When the 22d MAU first arrived in November, the HMM-261 commander, Lieutenant Colonel Granville "Granny" R. Amos, put two of his Cobras on the *Trenton*, fully armed and on a 30-minute alert. Cobra pilots and maintenance crews were rotated from the *Guam* every five days. A third armed Cobra was ready as backup on the *Guam*, while the squadron's fourth gunship was undergoing maintenance work. The

Cobras were never flown over the beach, but when the turnover took place in November, they were airborne, orbiting two miles off the beach. They were airborne again when the Embassy was evacuated in early 1984, and once again when the 22d MAU re-embarked in February. The gunships trained regularly, "running close air support with the fixed wing off the *Indy* and the *JFK* [the carriers *Independence* and *John F. Kennedy*], and they did a lot of naval gunfire exercises in case we did need them to go over the beach."[14]

In order to employ the helicopters usefully ashore, Major William J. Sublette, the MAU Air Liaison Officer, suggested a tasking for the squadron's other aircraft. Lieutenant Colonel Amos would launch a UH-1N, four CH-46s, and sometimes two CH-53Ds in the morning. These helicopters would then be turned over to Major Sublette's control ashore. At noontime, the planes would return to the *Guam* where new pilots and crews would take over. This was necessary because on some days, the pilots could fly for 8-10 hours without respite, "and that worked out really good as far as getting the max utilization of the airplanes and air crews without having a lot of dead time orbiting the airport."[15] If the aircraft were not needed at any time during the day, they would set down in one of the landing zones and shut down their engines. All helicopters returned to the *Guam* at night to avoid being hit by the random fire falling within the MAU perimeter. For medical evacuations, a CH-46 was also put on the *Trenton* on a 30-minute alert at night.

During the first three months into this deployment, HMM-261 fully supported the MAU with 25 percent of its flight time spent in ferrying VIPs and visiting entertainers around, and another 25 percent in supporting the Navy.[16]

One of the major characteristics of this deployment with respect to helicopter operations was the uncertainty from day to day about the security of the individual landing zones. The erratic and sporadic nature of the attacks on the helicopters was dramatized on 28 January 1984, when an unidentified individual fired a SA-7 missile at a CH-46 approaching a landing zone which had been used extensively since the 22d MAU's landing in November. Fortunately, the SA-7 missed the aircraft. Many flights had previously brought in external fuel loads here, hovering over the LZ without any problems. Earlier, on 8 January, another had flown into LZ Oriole, the landing area near the Embassy, which had been used without incident for two months. This time, however, several men fired upon the plane with small arms and RPGs, killing one Marine in the process.[17] In addition, there were many instances of helicopters flying to the beach and picking up indications that the aircraft were being tracked by a radar system that was associated with the Soviet quad-barrelled ZS-23mm antiaircraft gun. During the first two and a half to three months of the deployment, aircraft were constantly being tracked by radar as they flew into the airport. The HMM-261 helicopters were fired upon by small arms weapons, rocket-propelled grenades, and the single SA-7, but they received no ZS-23 fire.

When the squadron first arrived in Lebanon, it began averaging a total of 40 hours a day flight time. In December, this increased to 50 hours a day, with two or three days hitting 70 to 80 hours flight time. The squadron ended up the year with 1,415.8 hours of flight time in December, 1,348 in January, all the while averaging 90-95 percent aircraft availability. HMM-261's workload didn't lessen in February, for in 29 days, the pilots flew 1,417 hours for a 49-hour daily average.

All MAU components conducted on-the-job training when they could, in between times filling and hauling sandbags. By the first week of December, the Seabees had completed emplacing all sea-land containers in the MAU command post area. At the same time, the combat engineers attached to BLT 2/8 completed new fighting positions throughout the BLT area, and also emplaced barbed wire obstacles in front of each position.[18]

The week of 3-9 December was characterized by a series of violent clashes which resulted in the MAU's first casualties of the deployment. On 4 December, Navy jet bombers flew from the flight decks of the *Independence* and the *John F. Kennedy* to attack selected targets east of Beirut.* In response to anticipated retaliatory action, the MAU set a maximum alert condition, beginning at 0700.

During the course of the day, Marine positions on the eastern and southern airport perimeter were taken under occasional sniper and mortar fire, which was returned in kind. At 1935 and 2010, Checkpoint 7, a combat outpost located on Pepsi Road, which led towards the airport from Ash Shuwayfat past the Pepsi Cola bottling plant, was hit by small arms fire. Manning this outpost was a rifle squad reinforced by a machine gune team, a sniper team, and a LAAW team from the assault squad of Company G's weapons platoon.

*Two of the planes were shot down during this raid, with one Navy pilot killed and the second bailing out over Syrian-held territory. He was later returned to U.S. jurisdiction.

Photo by GySgt Dale S. Welker

A view of Amal-held buildings from Company G, BLT 2/8, positions. The arrow points to "Cafe Daniel," from which hostile small arms fire was directed against Marines.

The fighting positions of these Marines were atop two small, 2 1/2 story buildings, with each rooftop measuring 9'x12' at most. These positions were selected because they provided the best observation of all the small buildings in the area. An air assault company of the LAF 33d Battalion was collocated just southeast of the Marines.

At 2204, the eastern perimeter again came under fire, with one 122mm rocket round landing directly on top of Checkpoint 7, killing eight Marines and wounding two others. When the small arms fire directed at Checkpoint 7 increased, six off-duty Marines had rushed to the rooftop to reinforce the four already there. All became casualties. Of this tragic event, Lieutenant Colonel Smith later commented, "Good men rushed out of protective bunkers and into a fighting position. They felt it was the thing to do, and I don't fault them. I wish now they hadn't."[19]

Following this, the MAU returned fire with small arms, 81mm mortars, 155mm artillery, and 5-inch naval guns.[20] This response caused several secondary explosions, but enemy losses were unknown. The Marines could only tell whether they had inflicted casualties with return fire during daylight hours when they could see Lebanese civilian ambulances evacuating wounded. The BLT commander did not believe that the fire that killed his eight Marines was so-called spillover, ". . . I think from the very beginning that they were shooting at us . . . it is my opinion that it was because of the air strike that morning . . . . And there is no way of really documenting that the fire was because of the air strike."[21]

Two nights later, on 6 December, a short but violent firefight erupted near Company G positions. The fire came from fortified bunkers believed held by Amal radicals. After Marine small arms and machine gun fire, as well as M203 grenades, failed to silence the Amal, Marine tank and Dragon rounds finally did, destroying two bunkers in the process.

Relative quiet prevailed for the next few days, but then short, bitter firefights began in the early morning hours of 8-9 December, again in front of Company G positions and emanating from "Cafe Daniel,"[22] a known Amal position that had been fortified, and had firing slits directly facing the Marines.

Around this time, the Amal in Burj al Barajinah seemed to think that they had a special relationship with the Marines. On the evening of 6 December, several Amal appeared at the airport and complained to the LAF liaison officer that the Marines building bunkers on the eastern perimeter were impinging on Amal territory. They said, ". . . that it was too close to them and they wanted it stopped. If we didn't stop it, they were going to shoot at us. Well, we weren't building bunkers any further forward towards them than where they'd [the bunkers] always been."[23] Major Alfred L. Butler III, the MAU liaison officer to the Lebanese Army, quietly took notes while avoiding direct contact with the Amal.

Lieutenant Colonel Smith's response to this warning was that the Marines were only building defensive positions and clearing fields of fire. Further, since he had no offensive intent then, he said that he would continue to improve his defenses. On the morning of 7 December, while the Marine engineers worked with the Seabees in front of Company G positions, the Amal opened up with grenades, small arms, and machine gun fire. The Marines returned fire with tank rounds, Dragons, LAAWs, and M203s. After an hour, the firing ended.

That afternoon, Amal representatives again met with the LAF liaison officer and repeated their statement of the previous evening—that if the Marines kept working in front of Company G, they would be fired upon. True to their word, the Amal fired at the engineers and the Seabees during the morning of 8 December. This time, however, the Marines responded in more than kind, destroying all of the Amal bunkers to their front, including those in the "Cafe Daniel" building.

While this was going on, the Amal called the American Embassy to ask how they could arrange a ceasefire. They complained that the Marines weren't "responding in kind, that they thought they had an agreement. . . . Well, they didn't have any agreement, but that had been the rules of engagement, and they were aware of them, I guess."[24] Prior to this time, and certainly prior to the 23 October bombing, the rules of engagement decreed that Marines would respond proportionally to any life-threatening fire from any quarter. "Well, after 23 October, that made no sense."[25] And so the fire the Marines returned on 8 December was intense enough to destroy the positions firing upon them and lethal enough to cause Amal casualties.

On the morning of the 9th, the Americans suffered two more casualties. A Seabee was slightly wounded and his bulldozer just about destroyed when it was hit by an RPG. In the same attack, a Marine was shot in the leg and evacuated to the *Guam*.

After these incidents, things slowed down somewhat, but the Marines continued to receive fire from small arms and automatic weapons, and occasionally mortars. They were "obviously firing directly at us, and when we could determine where the fire was coming from, we responded, vigorously. Vigorous became the byword for our response. 'You shoot at us, you must be prepared to receive a vigorous response.'"[26]

During this time, the MAU continued to upgrade its positions, using the 33-man combat engineer platoon from Camp Lejeune that augmented the MAU's organic engineer capability in the BLT and the MSSG* and the Seabees' efforts. The Seabees were due to leave 23 December, and General Joy was determined to use them as fully as he could in their remaining time in Beirut. Meanwhile, Marines in Beirut continued to be visited by congressional delegations, as well as by high-ranking officers in the chain of command. In addition, General Joy was kept busy meeting with his MNF counterparts and with General Tannous.

By mid-December, 50 of the planned 80 sea-land containers had been emplaced as bunkers, with the remainder scheduled to be in place by the first of the year. Surprisingly, the weather continued to be fair, giving Marines an extra measure of time to work on improving fields of fire, building berms, and emplacing wire obstacles in front of their positions. Lieutenant Colonel Smith organized the BLT's defense along the eastern perimeter by pulling back to give Marines on the line as much open terrain—and as many good fields of fire—as possible. He then built platoon-sized strongpoints, ". . . really hardened . . . that, if it came down to defending against a major attack, each of these strongpoints could really fight and defend themselves. And that, of course, left . . . in

*See Appendix B.

Only a sandbagged post is visible on the skyline at Beirut International Airport.

Photo by GySgt Dale S. Welker

Golf [G] Company's case, as much as 300 meters between strongpoints."[27]

All of these strongpoints were covered by fire and observation. Lieutenant Colonel Smith left quite a few of the old, above-ground, "presence" bunkers in place, and at night he would send two- and three-man security patrols out to those old bunkers. The Marines would light up cigarettes, and occupy the bunker for several minutes, and then move out to another bunker down the line to do the same thing, to "let them [the Shiites or Amal] know they're in the bunker, and then let them always wonder where they [the Marines] are."[28]

In organizing the ground, the BLT moved around the terrain in front of the company positions. At the southern end of the perimeter, all of the ground in front of Company G was re-arranged by the Seabees and their bulldozers. According to Lieutenant Colonel Smith, he:

> . . . told Golf Company and Fox Company commanders to use their imagination and look at how they wished that terrain was, then make it that way. . . . There were several places over an Golf Company's frontage where over the years the Syrians had pushed up dirt facing south, Israelis came in and they pushed up the same dirt and reoriented it, but it was facing north. . . .So, those fire piles and artificial terrain . . . they turned it so it faced the way they wanted it to face. We moved a lot of dirt that way.[29]

A great deal of money was poured into these efforts to reorganize the defense. Based on an agreement between Generals Miller and Tannous—each acting for their respective governments—FMFLant would pay for all materials and costs for renting heavy equipment* employed in emplacement of the bunkers and building new fighting positions. Simultaneously, the Government of Lebanon agreed to underwrite the cost of installing lighting and placing concrete Dragon Teeth obstacles around the MAU perimeter.[30]

In the MAU's weekly situation report, General Joy advised that he planned to request the Marine Corps to put certain pieces of heavy equipment on standby for immediate airlift to Beirut, should the local supply no longer be available. He also noted that he was continuing his attempt to reduce the number of MAU personnel ashore and that in the second week of December, he managed to return more than 100 MSSG Marines to the ships.[31]

Concluding this report, General Joy noted that the threat of a conventional attack on the Marines remained an ever-present possibility.[32] At the same time, the terrorist threat remained probable, in light of several small incidents directed at the French MNF.

Watercolor by Maj John T. Dyer, Jr., USMCR (Ret)

On Watch, Christmas 1983

The MAU commander noted the heavy attack on Companies E and G, between 1630 and 1920 on 15 December. At that time, the firing then going on between the LAF and PSP in the vicinity of the Marines had spilled over into MAU positions. Approximately 20 mortar rounds detonated near or amidst the Marines and they were repeatedly fired upon by .50 caliber machine guns and ZS-23s. The MAU answered with 81mm mortars, tank guns, 155mm artillery, and naval gunfire. During the heavy firing, a PSP representative contacted the U.S. Embassy's political officer, asking how they could turn off the bombardment. He was told that if the PSP would stop shooting at Marines, they would not be fired upon. Shortly thereafter, Jumblatt's PSP ceased firing, while the Marines continued firing for 15 to 20 minutes more to ensure that all their targets were neutralized. General Joy wryly commented later, "It would appear our aggressive response to attacks by fire, and especially the *New Jersey*, has made an impression on some elements."[33]

*Because the heavy equipment needed to build these new defenses was not immediately available through military channels and not part of the equipment organic to the MAU or its component units, arrangements were made through the Government of Lebanon to rent them from local civilian construction firms.

Sketch by Cpl Michael Shelton

The 22d MAU's 1983 Christmas card reminiscent of the Iwo Jima flag raising. The tree is a Cedar of Lebanon and symbolizes the country in which it grows.

BLT 2/8 Chaplain Thomas W. Falkenthal, dressed in a Santa Claus suit, delivers gifts to Marines on the MAU perimeter from the rear of an ambulance. LtCol Edmund J. Connelly, Jr., 22d MAU G-3, is at the right.

Bob Hope, Miss America 1983 Debra Maffett, Ann Jillian, Kathy Lee Crosby, and Brooke Shields visit Beirut servicemen on board amphibious squadron ships at Christmas 1983.

As Christmas approached, the MAU was visited by a number of high-ranking military and civilian personnel. In December, the MAU and the Phibron were inundated with tons of mail addressed "To a Marine, 22d MAU," or "To a Sailor, Phibron 4." Included in the mail bags were countless fruit cakes, Christmas cards, thousands of pounds of cookies and candy, and the like. Lieutenant Colonel Amos, commander of HMM-261, recalled that someone sent three pallets of live Christmas trees to the American forces in Beirut, each tree decorated with ornaments. One could go to any one of the squadron's work shops ". . . any time from probably the 10th of December to the 10th of January and there would be five gallon pails of cookies that people had sent. . . . The mess decks [of the *Guam*] were plastered with 'Dear Sailor' Christmas cards. Just unbelieveable. Nobody could remember seeing anything like that. The outpouring of the American people. . . ."[34]

The phenomenon was not exactly new to Beirut Marines, for the 24th MAU had experienced something like this expression of American generosity when it was in Lebanon the Christmas before. Also, as the 22d MAU left Grenada, it received mail bags filled with cards, letters, and boxes of cookies and candy, thanking the Marines and sailors for what they had done in Operation Urgent Fury.

Carrying on a tradition he had begun in World War II of spending Christmas with U.S. forces overseas, Bob Hope and his troupe of entertainers arrived off Beirut just before Christmas to give shows on board the *Guam* and the *New Jersey*. Four hundred 22d MAU Marines were flown to the *Guam* on 23 December to see the show, while another 400 attended a show on the *New Jersey* the next day. Mr. Hope insisted on visiting the Marines who were not able to see his show, and on Christmas day he was flown ashore to make a quick visit to the MAU headquarters.

Also visiting the Marines during this holiday season, were Captain Eli Takesian, ChC, USN, Chaplain of the Marine Corps, and Captain Angelo J. Libera, ChC, USN, senior chaplain of the 2d Marine Division, who visited all MAU positions, holding Christmas services for the Marines ashore. On Christmas Day, Lieutenant Thomas W. Falkenthal, ChC, USN, the BLT chaplain—who had brought a Santa Claus costume with him to Beirut—donned it and went around to all of the BLT positions by ambulance, handing out Christmas presents to the line Marines.

By Christmas, 95 percent of the tank ditch around the MAU perimeter was completed, with 70 percent of the Dragon Teeth in place. The dirt berm around built-up area was completed, but only 20 percent of the planned wire obstacles were in place. Of the 156 planned fighting positions, 75 percent had been completed.[35] The new MAU command post was also sufficiently prepared to permit transfer of essential command/control/communications functions from the old, so-called "vital area" to the new position.[36] On 23 December, 29 of the Seabees sent to assist the MAU in building up its defenses were sent back to Rota. The remaining 12 were to stay in Lebanon for an additional 30 days.

During this Christmas period, the attacks by fire on Marine positions continued at a much-reduced level. A resupply convoy returning from the U.S./UK Embassy took fire with neither damage nor casualties. An Air Force bomb-dog handler in the Embassy area was slightly wounded by sniper fire on 22 December near the bombed-out American Embassy, while conducting a search for car bombs along Ambassador Bartholomew's usual automobile route near the temporary embassy site.[37]

By the end of the year, in unseasonably good weather, all major MAU command post functions were located in the new command post site. The MAU headquarters had been relocated on 27 December. New 9'x9'x13' living bunkers were completed for the Marines manning the amtracs on the Corniche near the joint embassy site, and three more were constructed in the rear of the Durrafourd Building for the guard platoon assigned to embassy security.

Ever since his arrival in Lebanon, General Joy had pressed for President Gemayel's approval to set up a Multi-National Force liaison/coordination office at the Ministry of Defense, primarily because the liaison office at the Presidential Palace was not operating effectively. The Lebanese officers at the Presidential Palace:

> . . . were a step behind the operational usefulness of the information that was passed to the MNF liaison officers. It was like a press debrief of the previous day's events and we didn't get anything in a timely manner or know exactly what was going on . . . in the detail or accuracy that was needed for tactical planning in defense of our forces and accomplishment of our mission."[38]

The problem was that General Tannous and his staff operated in the Ministry of Defense, where the action, planning, and timely information could be found. Seeing that General Tannous was unable to allow the overt establishment of an MNF functional coordination center at the MOD, General Joy and Lieutenant Colonel Rice, the 22d MAU's special staff officer, sought an opportunity to establish the function without formalizing it.[39] On 29 December, Lieutenant

Colonel Rice was assigned to duties at the Ministry of Defense with the Office of Military Cooperation to serve as a liaison officer between the Lebanese Armed Forces and General Joy, in the latter's function as Commander, Task Force 62, on matters concerning the United States Multi-National Force.[40]

General Joy commented that General Tannous gave this arrangement unofficial blessing. "However he requested that we maintain a low profile because of some reservations on the part of the Palace."[41] Lieutenant Colonel Rice functioned in this billet from 29 December to 24 February, after which he returned to the MAU headquarters and reassumed his assignment as MAU executive officer.

During his time at the Ministry of Defense, Lieutenant Colonel Rice frequently visited the LAF operations center, checking with Lebanese operations officers, intelligence representatives, fire support coordinators, and duty officers. As a result of these contacts, he was able to provide General Joy, and offshore naval units up-to-date target data. This ultimately meant U.S. Multi-National Force and its supporting arms could respond to General Tannous' requests in a more timely and suitable fashion.[42]

Meanwhile, the MAU continued its defensive construction efforts. By the first week of 1984, work at the U.S./UK Embassy and the Durrafourd Building was progressing satisfactorily. The prefabricated bunkers and fighting positions on the Corniche were completed. The sandbagging of a trailer in the rear of the Durrafourd Building and of a prefabicated protective bunker was almost completed with the assistance of a 20-man working party lifted by helicopter from the airport to the embassy area each day.

Serving as embassy guard for its entire period in Lebanon was Second Lieutenant Michael L. Ettore's 1st Platoon, Company F. He and his advance party were lifted by helicopter to the Embassy on 13 November, and the rest of his platoon joined him five days later. Although he was isolated from the rest of the MAU at its airport location, and was situated in the heart of Muslim-held territory in west Beirut, Ettore felt safer there than at the airport. Muslim factions were doing most of the shelling of the airport and he felt that the Muslims were not about to shell their own people.[43]

All of the posts of this embassy guard—not to be confused with the Marine Security Guard detachment inside the Embassy—were fortified bunkers in which the guards did tours of six hours on and six hours off. Initially, Lieutenant Ettore's detachment consisted of one officer, and 64 enlisted Marines, supported by three amphibious assault vehicles, two jeep, two Air Force bomb dogs and their handlers, a cook, and two Navy corpsmen, all of whom were reinforced by a rifle squad from Company F's 3d Platoon, a two-gun machine gun squad, and a squad from the company's weapons platoon.[44]

When things became hectic in west Beirut in early February 1984, the embassy guard was reinforced by another 35 Marines, approximately. While the guard was not fired upon purposely, it did receive some spillover fire and stray shots. From time to time, a phantom mortarman fired from never-discovered positions without causing Marine casualties. The rules of engagement for the guard changed somewhat from what they had been before the BLT bombing. When Lieutenant Ettore relieved the 24th MAU's Marines in November, he was told, "If there's a man on the roof . . . and he's got an RPG and he's obviously . . . going to shoot it at you, then you don't have to wait to be engaged because of the situation we were in. We didn't have the 400 or 500 meters buffer zone like they did at the airport." The Muslims were quite close to the Marines "and we could get shot at from 10 feet away."[45]

Surrounding the embassy area was a fairly large group of Druze PSP militia. They apparently had a good talking and working relationship with the American Embassy's Regional Security Officer, Alan O. Bigler, with whom the MAU Marines worked very closely. Having been in one position for so long, Ettore and his men were able to recognize individual PSP militiamen personally and at times were able to deal with them through Bigler. Once, when Ettore needed some dirt to fill sandbags, he passed the word to Bigler, who, in turn, told Salim, the local PSP leader, "and the Druze actually hauled us in some dirt."[46]

Despite the heavy fighting which erupted in west Beirut in early February, the *status quo* between the Marines and the PSP remained in force, and the Marines were not fired upon by the locals. Several unknown assailants did, however, fire upon Marines unloading a helicopter at Landing Zone Oriole, near the embassy, without causing any casualties. According to Ettore, Salim told him that they were not his men, and that "several times, when some of his people caused incidents, he would just simply offer to kill them to show his sincerity. He said, 'Do you want me to kill them?' And I would say, 'No, no!' But all you had to do was just tell him, 'Look, this guy is bothering us, don't let him back here,' and you'd never see the guy again."[47]

Meanwhile, events beyond Lebanon were beginning to determine the future of the Marines in that country. Some segments in American politics and society were completely opposed to the deployment of Marines to Lebanon and the nature of their mission. They pronounced dire forebodings of what would result from the continued presence of Marines in this troubled area of the Middle East. The quickest way to get the Marines out was for Congress to invoke the War Powers Resolution.* After much heated debate, the Congress granted the President authorization to keep the Marines in Lebanon for 18 months.

Following the bombing of the BLT headquarters, publication of the Long Commission and House Armed Services Committee reports, and a period of public mourning, there was increased pressure upon the Administration to pull the Marines out of Lebanon, an action the President adamantly refused to take. None of this clamor in the United States escaped the notice of the Beirut Marines. The media, for example, were constantly asking Marines what they thought about the Beirut situation and how they felt about remaining in Lebanon. Digging in at the airport and witnessing the increased strength and effectiveness of the various militia factions surrounding their positions (as well as the inability of the Lebanese Armed Forces to impose its will on the government's enemies), the Marines began to realize that perhaps their time in Lebanon was growing short.

Rumors—"scuttlebutt" to the Marines—began to circulate within the MAU. As General Joy recalled, "I think Congress came back in session around the 23d of January and we had a constant stream of Senators and Congressmen coming to visit us all during December and January. And it was very obvious that there was going to be a big battle on the floor of Congress over getting the Marines out of Beirut."[48]

In the midst of all this, the MAU continued to improve its positions and to respond to those who fired on them. In the evening hours of 7 January, after a lone rifleman fired on Marine positions at the northeast perimeter of the airport, the Marines there returned aimed rifle fire and one M203 grenade ". . . which blew the attacker out of sight."[49] At about the same time, a 107mm rocket impacted near Marine positions in the southeast perimeter, wounding two Marines. Apparently, this was spillover fire coming from LAF-Druze fighting nearby.[50]

A Marine was killed on 8 January while on a work detail at the Bain Militaire on the Corniche near the U.S./UK Embassy. Five days later, while improving positions at the southern end of the airport, Marine combat engineers were fired upon by unidentified individuals from a building nearby, known locally as the "York Building." There were no Marine casualties, but the Marines returned a "decisive volume" of small arms, mortar, tank, Dragon, and LAAW fire which severely damaged the building and quieted the hostile fire.

By 12 January, only 31 Marines remained in the old MAU CP site, and the new MAU headquarters was in full operation. Phase I of a three-phase construction program had just about been completed. By 16 January, in planned Phase I construction, 119 sea-land containers had been prepared and 130 emplaced. Thirty-two prefabricated bunkers had been completed and 45 emplaced. With respect to fighting positions, 156 had been prepared, 128 completed, and 136 emplaced.[51] The cost of material, of equipment rented from local sources, and civilian labor came to $1.56 million.[52]

Meanwhile, Phase II construction went on. In this phase, the MAU planned to construct protected bunkers for all personnel ashore. Material for this construction continued to arrive in Beirut. It arrived either by ship, usually the USS *Transcolumbia*, or by helicopter from Larnaca. A major effort was also underway to complete the barrier plan, which included the installation of Dragon Teeth and a tank ditch all the way around the perimeter. By the middle of January, the Government of Lebanon's promise to install perimeter lighting was still unfulfilled. Phase II, when completed, would cost $771,000.

Phase III called for the reinforcement and hardening of all sea-land container bunkers to enable them to withstand direct hits from fuzed delay ordnance. The costs for 540 metric tons of steel I-beams, concrete, cyclone fence, waterproofing, lumber, nails, rented equipment, and civilian labor would total $3.705 million.

The actual construction for all phases was done by 74 Seabees and 99 Marine combat engineers. All told, they emplaced more than 400 sea-land containers, 192 bunkers, and 156 two-man fighting holes.

In addition to this three-phase construction effort,

---

*Essentially, the War Powers Resolution states, among other things, that if U.S. forces are introduced into hostilities or a place where hostilities are imminent, the President will report to the Congress within 48 hours of taking such action the circumstances necessitating this action, the constitutional and legislative authority on which such action was based, and the anticipated scope and direction of the hostilities. From this time, whenever the President has reported to Congress that he has taken such action or plans to, he has 60 days to recall the forces, unless (1) Congress declares war or authorizes such use of force; (2) Congress extends the 60 days period; or (3) Congress is unable to meet because of an attack on the United States.

Shown in this photograph is one part of the Dragon's Teeth perimeter surrounding the 22d MAU's positions at Beirut International Airport. Nearly 800 were emplaced.

A view of the road leading north to Beirut from the airport in January 1984. Note that a berm has been constructed and Dragon's Teeth are in place outside the MAU compound. To the right is the devastated BLT building and in the background is Beirut itself.

Photo courtesy of BGen Jim R. Joy, USMC

the MAU implemented a barrier plan. In building the barrier during the period beginning November 1983 to January 1984, the MAU accomplished the following:

a. Constructed and strategically emplaced 500 concrete Dragon Teeth;
b. Constructed a 9-foot-high dirt berm around the MAU perimeter;
c. Set the existing steel fence along the coastal highway in concrete;
d. Constructed a tank ditch around the MAU perimeter;
e. Strategically placed a double apron, triple concertina wire barrier around the perimeter of MAU positions;
f. Strategically placed trip flares and "flash bangs" around the perimeter;
g. Employed derelict vehicles as obstacles along the perimeter;
h. Constructed tetrahedrons for placement at the gates leading into the MAU positions to slow down vehicular traffic.[53]

When the MAU received orders to leave Lebanon, consideration was given to destroying the bunkers and all defensive positions. However, it was determined that such an effort would be too costly and time-consuming.

Near the end of January, the MAU was furnished six M19 40mm machine guns, .50 caliber sniper rifles, improved 60mm mortar rounds, and additional night vision goggles.

During the evening of 14 January, the MAU's eastern perimeter positions came under small arms fire, but no casualties resulted. Later that evening, positions on the southern perimeter received four large-caliber rounds, which caused no damage. After the Marines fired three 155mm illumination rounds at suspected firing positions, firing ceased.

The next night, the perimeter came under a large volume of fire of all calibers from the hostile firing positions running along a ridgeline east of the airport. A 122mm rocket hit the Marine fuel farm inside the perimeter destroying large fuel bladders and igniting 2,500 gallons of gasoline. The Marines returned fire vigorously, calling in 5-inch naval gunfire. After the firing ceased, there were no Marine casualties. The number of enemy casualties was unknown.

Toward the end of January, MAU representatives met with embassy personnel and the staff of Phibron 4, to review contingency plans for a non-combatant evacuation operation. Such an operation was not actually contemplated at the time, but the timing for the review would prove to be near perfect.

To break the routine of filling sandbags and improving their positions, the Marines conducted on-the-job training, held classes in first aid, and trained in the use of TOW/MULE (Modular Universal Laser Equipment) night sights.*

Concerned by the possibility of kamikaze air attacks on MAU positions and Phibron shipping, FMFLant sent an air defense survey team to Beirut to assess the air defense requirements for the airport, the U.S./UK Embassy, and the Durrafourd Building. The team developed a defensive concept that called for an additional 39 Marines. General Joy noted that this was an unacceptable number in view of his efforts to reduce the number of men ashore, ". . . but appeared warranted to provide a viable air defense/control system."[54] On 10 February, the MAU was augmented by ten Stinger Missile Teams from the 2d Marine Aircraft Wing. Six were deployed at the airport, two at the U.S./UK Embassy, and two held in reserve.[55]

At about 0830 on 28 January, a SA-7 missile was fired from a position northeast of the airport at a CH-46 helicopter landing in the vital area. The missile missed its target and landed in the sea. Two days later, Amal elements in the vicinity of Cafe "Daniel" fired small arms and rocket propelled grenades at Company G positions killing one Marine and wounding another. The Marines responded once again with tank gun fire, 60mm mortars, M203 grenades, fire from 40mm and .50 caliber machine guns, and small arms fire. This resulted in an estimated three Amal killed and 11 wounded.[56]

The firing continued throughout the day of 30 January, escalated in mid-afternoon, and finally ended approximately three hours later. For the first time during this deployment of the 22d MAU, the vital area (former location of the MAU headquarters), was hit by 15-20 mortar rounds. One Marine was wounded. Company E, on the perimeter, was also hit by seven more mortar rounds. At about the same time, a Company G radioman was hit and killed by a sniper.

The source of the fire was located by the Marine Target Acquisitions Battery (TAB) attached to the MAU and the Army TAB, similarly assigned, but the Amal mortars were firing from heavily populated areas. Under the existing rules of engagement, the Marines were prohibited from firing on areas where there would undoubtedly be "significant collateral damage" (e.g., civilian casualties). General Joy was able to fix one position in a graveyard, and passed target information about this and a second position to the LAF with the request that they place fire on them. The LAF complied, but other Amal positions were not so easily

*The MULE proved invaluable in determining the accurate ranges of targets and key terrain features, and in the designation of targets for aircraft acquisition and engagement.

reached. Some positions were located so far north that the TAB radar fan could not cover them.

The MAU commander noted in his situation report for the period, "The most troublesome matter is that two Marines, well protected, ended up as casualties. This is a very sobering point, that readily demonstrates the fact that we may take casualties regardless of how well we defend ourselves."[57]

At the end of the month, tension was visibly rising between the LAF and the militia of the various factions. Rumors spread throughout Beirut and the suburbs of an impending major government operation against the militia, and all parties prepared for even heavier fighting. It appeared to the BLT commander that the Amal and PSP well knew LAF plans ahead of time, and had begun an offensive of their own against the Army before the LAF could begin theirs.[58] The fighting was particularly heavy on the night of 4 February, with some spillover fire hitting inside the Marine perimeter. Especially heavy fighting broke out between the LAF and Muslim elements in Beirut and the southern suburbs on 5 February. In the late morning, the Marine positions in the northeast portion of the perimeter were hit by both direct and spillover fire, but no casualties were sustained. That night, the LAF pounded the southern suburbs and Khaldah with tank main gun, artillery, mortar, rocket, and small arms fire. A backlash resulted from this heavyhanded effort when LAF Muslim soldiers, whose families lived in these areas, refused to continue fighting. Some left their units, while others just remained in their barracks. Meanwhile, Nabih Berri called the Amal out of the Lebanese Armed Forces, ". . . in fact, he called all Muslims to leave the LAF."[59]

The Marines could see what was happening in the LAF units closest to MAU lines. On the night of 5 February, the Amal and the PSP went on the offensive all over west Beirut and the southern suburbs. LAF units along the airport road leading from Beirut to the terminal essentially laid down their arms and left quietly, with the Amal just as quietly taking over the abandoned posts and terminal area that night. The only building they did not occupy housed the LAF liaison office. An Amal leader, Dr. Salinas, visited the office, ". . . and asked that the Marines be advised that 'the Amal does not want to fight the Marines.'" He reportedly requested that the Marines not fire on the Amal, and said, "Even if the Marines attack us, we will not return the fire."[60]

The LAF units east of Company G had a particularly hard fight that night. It lasted from about dusk to about 2230 before it died down, observed closely by the Marines. In front of the MAU positions were a Lebanese infantry company (reinforced by a tank platoon) and an air assault company. The LAF units had shared a checkpoint with the Marines on Pepsi Road. A telephone line went back to the Marine company command post. The LAF captain called Company G commander, Captain Robert K. Dobson, Jr., to tell him that the government troops still held all of their positions. At about 2300, loudspeakers in front of the LAF units began to blare messages in Arabic. By dawn the next day, the LAF commander had but few troops left, all of them Christian. His Muslim soldiers all had deserted. The Lebanese officer told Captain Dobson that he had to withdraw through the Marine lines because he only had about one-fourth of his former command remaining. Lieutenant Colonel Smith ordered Company E, less a few Marines holding their former positions, to fill in where the LAF companies had formerly been. At this time the Amal pulled back, indicating once more that they had no desire to fight Marine forces.

About 1530 on 6 February, a heavy volume of large caliber and small arms fire, originating from Druze-controlled areas, fell on MAU positions along the eastern perimeter. The Marines answered with fire from all their organic weapons, plus 5-inch naval gunfire. The MAU also called in the first Marine-controlled tactical air mission since the August 1982 landing in Lebanon. Directed by a BLT 2/8 forward air controller, a Navy A-6 Intruder from the carrier *John F. Kennedy* dropped two laser-guided bombs on an identified target. At 2230, firing on the Marines ceased. One Marine had been killed.[61]

The next day, LAF security around the airport* deteriorated at a rapid rate, as Lebanese soldiers, with their tanks and other rolling stock, sought a safe haven within U.S. positions at the airport, or continued on to the north to join up with other government forces. An hour after noon on the 7th, large caliber fire landed in the center of the airport, and 50 minutes later, the MAU evacuated approximately 250 personnel, including Seabees, Marine combat engineers, and other Marines to Phibron shipping. All construction work at the airport ended. General Joy planned to bring some of the Seabees back ashore, when possible, to finish emplacing the sea-land vans, but this plan was overtaken by events.

On 7 February, the MAU began non-combatant evaucation operations, bringing out 40 American civilian embassy employees and their dependents by

*Since September 1982, the Government of Lebanon's Army was responsible for the exterior security of the Multi-National Force units.

Civilians leave the Durrafourd Building for helicopters carrying them out of Beirut in February 1984.

Marines adjust a protective helmet on a youngster before he is evacuated from Beirut in February 1984.

helicopter from the evacuation control center at the U.S./UK Embassy to the *Manitowoc*. The next day, 49 more Americans were evacuated to the *Guam* for the airlift to Larnaca later. By 11 February, a total of 787 individuals had been flown from the embassy area or from Juniyah to Phibron 4 shipping, and then on to Larnaca.[62] On 11 February, one evacuee was hit in the neck by a stray bullet, but suffered only minor injuries. She was flown to the *Guam* for treatment.

Playing a major role in these evacuation operations was First Lieutenant Ettore's platoon from Company F. Ever since its arrival in Lebanon in November 1983, it had been providing security for the U.S./UK Embassy and Lieutenant Ettore had worked very closely with State Department representatives on the evacuation plan.

Heavy shelling in east Beirut was coming close to the Lebanese Presidential Palace, Ambassador Bartholomew's residence, and the Ministry of Defense on 8 and 9 February. The Government of Lebanon requested American fire support to engage the hostile artillery positions. Target acquisition units located the positions inside Syrian controlled territory. The Lebanese request was passed to higher headquarters for approval and once it was received, the *New Jersey* and the *Moosebrugger* took the positions under fire, silencing them.[63]

Plans for the withdrawal of the MAUs had existed since August 1982. When the 22d MAU landed in November 1983, the concept of redeployment was rediscussed. It became apparent to MAU staff officers as they read the message traffic through early February, that the Marines would be redeployed, but not

HMM-261 CH-46s lift off from the Corniche near the British Embassy in February 1984, evacuating civilians from strife-torn Beirut, when the situation became critical.

all at once. From the very beginning of his command in Lebanon, General Joy had been under pressure to reduce the number of Marines ashore. The MAU developed numerous plans to shrink the size of shore-based units while maintaining enough force to carry out its mission. According to MAU Assistant G-3, Lieutenant Colonel Van Huss, ". . . that was a continuous effort and a priority with General Joy and his staff and the commanders."[64]

It was also planned that—sooner or later—the Marines would totally seabase their logistical effort, leaving only a small combat service support detachment ashore. Plans for redeployment had been discussed before the LAF situation had deteriorated, ". . . and with the events of early February, it was prudent that we continue with [them]. Not in haste. It was programmed. . . ."[65]

Early in February, General Joy learned from Lieutenant Colonel Peter E. Woolley, commander of the British MNF contingent, that President Reagan had informed the governments of Great Britain, France, and Italy, that the United States was going to withdraw its forces from Lebanon. General Joy learned of the announced decision while listening to a British Broadcasting Corporation shortwave news broadcast on 7 February. The report stated that the President had ordered the Marines in the Beirut area to begin a phased withdrawal to Navy ships offshore shortly. Official orders had not yet reached General Joy.

This same day, the British contingent departed. Lieutenant Colonel Woolley called General Joy to tell him that he had received his marching orders and was leaving immediately. As General Joy recalled, Woolley said, ". . . 'I'm going to see General Tannous and tell him I'm leaving. We are going to motor march to Juniyah and will be picked up in Juniyah and leave.' And sure enough, they did."[66]

On or about 15 February, General Joy sent a message to the CinCEur planners stating that the MAU could pull out by 28 February if a redeployment was being considered. The MAU plan for a 28 February departure date provided for the possibility of up to two days of foul weather which meant that the Marines could actually leave on the 26th, the weather and other factors permitting.[67] This eventually became the day that the Marines left Lebanese soil. The JCS order to the MAU to execute the redeployment was sent on 18 February.

On 16 February, in response to the MAU message, General Rogers directed General Joy to turn over command of the 22d MAU to Colonel Faulkner on 20 February, and to establish and assume command of Joint Task Force, Lebanon (JTFL).[68] At the same time, the MAU was ordered to occupy and defend positions in the vicinity of Beirut International Airport—a MAU mission since September 1982—and to conduct a tactical reembarkation. The MAU was also directed to provide external security for the U.S./UK Embassy, and to support JTFL.

General Joy's new command would be comprised of the MAU; the Office of Military Cooperation; the U.S. Army Training unit located at the Ministry of Defense; and an embassy security detachment, made up of MAU Marines, responsible for guarding the U.S./UK Embassy and the American ambassador's residence. General Joy was further directed to maintain his command post at the airport until the MAU departed, and then to move it into a secure location in east Beirut. Since he would be working with the Lebanese Armed Forces, General Joy decided to set up his office at the Ministry of Defense.[69]

Once the MAU had re-embarked on Phibron 4 shipping, the Marines reverted to the operational control of the Sixth Fleet. General Joy had the 22d MAU under his JTFL command 20 through 26 February. Thereafter, he had only Lieutenant Ettore and 100 Marines who guarded the embassy, and 200-300 Army trainers in the Office of Military Cooperation, which consisted of three Special Forces training teams, each consisting of approximately 75 soldiers. General Joy also had an ANGLICO team to help carry out his fire support mission. He placed sections of this team in strategic vantage points in the mountains overlooking the city of Beirut and the Ministry of Defense.

The MAU was experienced in rapid re-embarkation, but the Marines had accumulated a large amount of excess gear over their 18 months' stay in Lebanon. The situation in Beirut prevented loading the *Transcolumbia* from the port, so the MAU's surplus supplies and equipment were loaded aboard the *Manitowoc* and the *Barnstable County*. The two LSTs then steamed to Haifa. After they docked there, the excess was transferred to the *Transcolumbia*. The LSTs then returned to Beirut, ready to begin a phased re-embarkation.

Throughout early February, fire had fallen sporadically on and around MAU positions, and the Marines continued to return fire. On 9 February, the Government of Lebanon requested naval gunfire placed on rocket positions which were firing on Beirut. The Navy complied with the request, hit the targets, and the firing stopped. The next day, Marine positions in the southern sector received heavy mortar fire, which was answered in kind by 60mm and 81mm mortars again silencing the enemy. However, three large caliber rounds exploded in the southern sector of the airport, severely damaging the counterbattery radar equipment supporting the Marines there, and effectively limiting their ability to determine where hostile fire was coming from.

On 14 February, the Marines were fired upon again, but suffered neither material damage nor human casualties. Meanwhile, General Joy discussed the situation with his French and Italian counterparts, and consulted with Commodore Erie and Admiral Martin, Sixth Fleet commander, about the future of the U.S. elements of the Multi-National Force in Lebanon. He continued to backload all non-essential personnel and equipment on Phibron shipping, and to refine re-embarkation plans.

In a report to General Rogers, the MAU commander stated that fire support coordination procedures had been simplified under new of rules of engagement. All U.S. elements could now respond immediately and directly to LAF requests for fire support. General Joy also reported the impending departure on 19 February of General Angioni's Italian units, with the last of them scheduled to leave Beirut on the 21st. The Italians were to leave behind at the port area a 100-man airborne company, while the San Marcos Tactical Group would remain on ships offshore, ready to land in a contingency.[70]

Even before the arrival of orders directing the MAU to leave Lebanon's shores, the MAU Service Support Group began backloading equipment, supplies, and personnel to comply with General Joy's directive to reduce the size of the MAU ashore. Up to this time, the MSSG was ". . . kept busy 18, perhaps 20, hours a day, in some cases [with] primarily what I just call routine support to the MAU; that is, maintaining the MAU with rations, with water, with fuel, with ammunition, all these other kinds of services . . . .[71]

When not busy with these jobs, the MSSG Marines were building the berms and digging the tank ditch, or filling sandbags. ". . . there wasn't a lot of free time, and there wasn't any place to go, so we stayed right on the beach and turned to."[72] On 13 February, Lieutenant Colonel Davidson, the MSSG commander, began backloading the MAU's Class I (rations), III (petroleum, oil, and lubricants), IV (construction materials), and IX (parts, repair kits and components) supplies to amphibious shipping. The next day, the remaining Seabees went aboard the *Transcolumbia* together with 22d MAU equipment. By 16 February, the MSSG had completed backloading excess supplies and had begun a phased redeployment of its personnel. Two days later, with nearly all supplies and equipment back on board ships, the MSSG commander established a combat service support detachment at the airport to support 22d MAU elements still ashore. But from this point, combat service for the MAU was

With spirits high and the U.S. flag waving, BLT 2/8 Marines—among the last to leave Beirut International Airport on 26 February 1984—wade through the surf of Green Beach to board landing craft which will carry them to Phibron shipping offshore and on to Rota.

Photo courtesy of Claude Salhani

Marines drive their vehicles into the well deck of the Barnstable County *(LST 1197) in February 1984.*

essentially seabased.[73] Lieutenant Colonel Davidson moved his command post on board the *Trenton* on 20 February, a full six days before the rest of the MAU boarded its ships.

The BLT began backloading on 9 February, when support elements and equipment began to leave the airport. The battalion's Headquarters and Service and Weapons Companies went aboard their assigned ships on the 25th, and the rest of the BLT left the airport the next day. At 0400 on 26 February, Company E was flown out from LZ Brown, near the terminal area and the north-south runway of the airport. Helicopters then returned to the airport for Company F. Both companies were back aboard the ship by dawn. Company G was slated to leave from Green Beach in armored amphibian vehicles and Phibron landing craft. The withdrawal of the BLT would have been completed by 0630, had not the Phibron's LCUs been given another task—the transporting of ammunition from Sidon to Juniyah—before loading the Marines at Green Beach. The last elements of the BLT left the beach at about 1237.[74]

CH-46 Sea Knights flown by HMM-261 pilots ("The Bulls") pass over Green Beach carrying Marines back to Amphibious Squadron 4 shipping offshore as the 22d MAU leaves the soil of Lebanon on 26 February 1984. Beyond the haze in the background is Beirut.

Photo courtesy of Claude Salhani

Lieutenant Colonel Amos, HMM-261 commander, recalled that the 26th was:

> . . . a beautiful Sunday morning. I remember, I flew back in and landed beside General Joy's Huey and sat up on the hill, you know, where the artillery positions were. And he and I and Ray Smith (BLT commander) sat there watching the beachmasters leave, birds were singing. . . . And you could see young kids moving in, playing in the bunkers down there to the north where we had moved out of. And [then we] flew back to the boat.[80]

Earlier that morning, another key event took place. The CinCEur directive of 19 February had ordered General Joy, to "effect liaison with Lebanese government to ensure that security for the airport was turned over to the Government of Lebanon."[76] At that time, however, neither the Lebanese Government nor its army had a responsible individual or unit at the airport or in its proximity with authority to accept responsibility for airport security.

After the heavy fighting of 7-8 February, when the LAF's 4th Brigade left the southern area of the airport and gave up Khaldah, the Amal took control. One of the Amal representatives, a man named Tylass, who was described as a young Muslim war chief, contacted the Americans and said, "We are responsible for west Beirut, we are responsible for the southern suburbs," and "we will see to it that the airport is safe, we will see to it that the Marines are not attacked, we will ensure that only authorized vehicles will transit the coastal highway." The Amal did what they promised to do.[77]

At 0600 on 26 February, control of Beirut International Airport was turned over to Captain Habib, representing the LAF 33d Battalion, 3d Brigade.[78] Shortly thereafter, Colonel Faulkner, Lieutenant Colonel Van Huss, and Major William J. Sublette, the MAU Air Liaison Officer, went to the LAF liaison office at the airport to recover the American flag, which had been there for some time. They had planned to bring the flag back to the States to present it to the widow of Major Alfred L. Butler III, the MAU liaison officer to the LAF, who died as the result of an accidental discharge on 8 February—the last Marine to die in Lebanon.

As Lieutenant Colonel Van Huss recalled the scene:

> . . . Colonel Faulkner turned to Colonel [Fahim] Qortabawi [the LAF liaison officer], and said 'With your permission, we will now strike our colors.' Bill Sublette and I moved immediately to the flag staff, took the flag down, folded it properly as it should be, and as we were folding it into the triangle, Colonel Qortabawi, perhaps was a little bit taken by the seriousness of what we had been doing. He reached up and took the Lebanese flag down, folded it—I don't know if he folded it properly. . . . He simply folded it and handed it to Colonel Faulkner and said, 'Well, you may as well take our flag, too.' And it was over.[79]

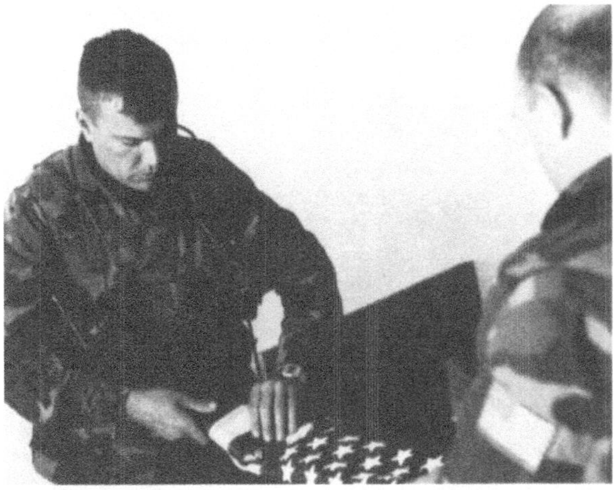

Maj William J. Sublette, 22d MAU Air Officer (left), and LtCol Earnest A. Van Huss, 22d MAU Operations Officer, carefully fold the American flag which hung in the Lebanese Armed Forces airport liaison office.

As though he really didn't fully understand the significance of the moment, Colonel Qortabawi said to Colonel Faulkner, "You are leaving?" The MAU commander replied, "Yes, we are really leaving. Our eastern positions have already been vacated, we're in pullback positions now, holding in the vicinity of the high ground down near where Hotel Battery was emplaced [on the western edge of the airfield], and we are in the final throes of embarkation. Yes, Colonel Qortabawi, we are really leaving." Again, Lieutenant Colonel Van Huss recalls:

> Colonel Qortabawi was a Christian. He said, "I have no way to go home. To go home, I have to go through Muslim checkpoints. You can get me to the Ministry of Defense by helo ride?" [Col Faulkner replied] "Yes, we can do that." So Colonel Qortabawi left with us; we gave him a helo ride to the Ministry of Defense, he linked back up to General Tannous, and it was all very final and over.[80]

The 22d Marine Amphibious Unit left behind more than one million filled sandbags and a lot of deep holes, which the Shiite militia Amal very quickly occupied.* The Marines departed with all that they had brought with them, leaving behind very little in the way of scrap materials. There was some thought of cratering the emplacements that they had dug and destroying the sea-land vans, but, in the minds of the MAU's staff officers, it had taken a Herculean effort

*"Exactly six minutes after the last [Marine] amtrac left [the beach], the Amal flag was flying over the watchtower at Black Beach. Likewise, Amal flags were going up all over the airport." Larry Pintak ltr to author, dtd 10Jan87.

Photo courtesy of Claude Salhani

Security men with weapons at the ready surround Secretary of Defense Caspar Weinberger, in helmet and flak jacket, as he holds an impromptu press conference before meeting with U.S. Ambassador to Lebanon Bartholomew at the latter's residence.

to get them in the ground, and it would have taken a similar effort to have dug them out. The Dragon Teeth were left where they had been placed, for only a heavy crane could have lifted them.

The MAU remained on board its ships until relieved on 10 April by the 24th MAU, commanded by Colonel Myron C. Harrington, Jr. At that time, the new amphibious task force took position hull down on the horizon, just out of sight of the Beirut shoreline. On 29 February, the Secretary of Defense visited the 22d MAU and Phibron 4 and presented the Navy Unit Commendation to both commands. General Kelley visited his Marines on 11 March.

The MAU still had a role to play in Lebanon—as a reaction force to rescue the American ambassador, if necessary, or in other contingency operations in Lebanon or elsewhere in the Mediterranean. Meanwhile, Phibron 4 ships, with embarked Marines, would leave, one at a time, for port calls at Haifa and liberty for all hands. During one such port call, a young Marine was killed in Haifa in an automobile accident. He was the last MAU Marine to die while the 22d MAU was deployed.

Marines lower the national colors for the last time in Lebanon at their Beirut International Airport outpost.

USMC photo by Sgt Hartman T. Slate

The turnover between the 22d and 24th MAUs was completed by 1000 on 10 April. At 1600, Commodore Erie's Amphibious Squadron 4 left the Lebanese littoral for Rota, where the MAU would wash down all its equipment. This task took three days, 16-19 April, after which Phibron 4 headed for the United States. Arriving at Morehead City on 1 May, the 22d MAU was given the same type of greeting by bands, cheeleaders, and officials which had met all the other returning MAUs. On 3 May the Secretary of the Navy and the Commandant reviewed the MAU. Eight days later, on 11 May, the 22d Marine Amphibious Unit was deactivated as Landing Force Sixth Fleet 1-84.[81]

The departure of the 22d MAU did not end the Marine presence in Lebanon. Still ashore were General Joy and his joint task force, the 24th MAU Marines who provided external security for the U.S./UK Embassy, and the Marine Security Guards providing internal security for the U.S./UK Embassy. Initially, General Joy was scheduled to deactivate the task force when the 24th MAU relieved the 22d. However, General Lawson, Deputy CinCEur, directed Joy to remain after the new MAU arrived to assist in getting it settled and to ensure that the 24th MAU and Phibron 6 instituted good working relations with Ambassador Bartholomew and the Office of Military Cooperation, as well as with the Lebanese.

The first orders General Joy received set a date of not later than 15 April for the deactivation of the joint task force. Meanwhile, a senior U.S. Embassy official was kidnapped in Beirut by an unidentified faction and General Joy was called upon to assist in the efforts to recover him. As a result, the deactivation was delayed. On 19 April, he was finally ordered to deactivate the force two days later and to inform Ambassador Bartholomew and General Tannous of his orders. The Ambassador was none too happy with this news nor with the timing for the deactivation. Peace talks were then being held in Damascus, and the American diplomat believed that President Gemayel was going to return to his capital and announce an accommodation with the Syrians. Ambassador Bartholomew felt, accordingly, that the deactivation was premature. He suggested that General Joy request a delay for several days. CinCEur concurred, and the Joint Task Force, Lebanon was officially deactivated on 26 April. General Joy flew to Stuttgart for a debriefing and then returned to Camp Lejeune, where he resumed his duties as Assistant Division Commander, 2d Marine Division.[82]

The final curtain came down on Marine Corps presence in Lebanon on 31 July 1984, when 24th MAU Marines guarding the U.S./UK Embassy returned to Phibron 6 shipping by helicopter and amtracs. The

The Beirut Memorial, at the entrance of Camp Johnson, Jacksonville, North Carolina, was dedicated on 23 October 1986. The concept for the memorial came from Jacksonville citizens and was paid for by donations coming from all over the country and world.

USMC photo by Cpl Jamee Sosa

last of the Marines departed at 1824 local time. This departure coincided with the transfer of American diplomats from the British Embassy to new offices in east Beirut, where space had been found to house the U.S. Embassy.

With the withdrawal of the last MAU Marines, Marine presence in Lebanon ended as quietly as it had begun spectacularly. The only Marines now remaining in Beirut are those assigned to the security guard inside the American Embassy.

For a time after the Marines left, there was an uneasy truce in Beirut. The Green Line was bulldozed out of existence and there were few reports of firing between factions. It almost seemed as though the peace that was elusive when the Marines were in country was almost within grasp. But it was never to be, for the fighting soon became as intense as ever.

In the 18 months that the MAUs were in Lebanon, 238 Marines died and 151 were wounded. Another 40 Marines suffered non-battle injuries, and seven were wounded as the result of the accidental discharge of weapons.*

To memorialize the Marines who served and died in Beirut, "Lebanon" was added to the battle honors of the Marine Corps already enscribed on the base of the Marine Corps Memorial in Arlington, Virginia. It was, perhaps, appropriate that this new battle honor was unveiled on 8 November 1985, when the traditional ceremonies celebrating the 210th birthday of the Marine Corps were observed.

A further memorial to the Beirut Marines was dedicated on 23 October 1986 at the entrance to Camp Johnson—the old Montford Point Camp—in Jacksonville, North Carolina. The concept of this memorial came from Jacksonville citizens, whose enthusiasm and dedication served to inspire donations from individuals and organizations from all over the country. Present at the dedication were the families and friends of those who died in Beirut and Generals Kelley and Gray, as well as now-retired General Miller, Colonel Geraghty, and Lieutenant Colonel Gerlach. The simple memorial consists primarily of two large walls. On the left side is inscribed the names of the soldiers, sailors, and Marines killed in Beirut and Grenada, while on the right wall are the words, "They Came in Peace."

Of these Americans, a proud but saddened and grateful Marine Corps and nation can only say, "Thank you" and "Semper Fidelis!"

*These statistics were obtained during a telecon between Hd, Casualty Section, HQMC and the author on 3Aug84. See Appendix F for a listing of Marine casualties in Beirut.

Notes

Chapter 1
Footnotes

1. The material for most of this section was derived from *Report of the DOD Commission on Beirut International Airport Terrorist Act, October 23, 1983*, dtd 20Dec83, pp. 94-98, hereafter *Long Commission Rpt*.
2. SSgt Randy Gaddo, "Beirut Bombing," *Leatherneck*, Feb84, p. 16.

Chapter 2
Footnotes

1. For a more comprehensive account of the 1958 deployments, see Jack Shulimson, *Marines in Lebanon, 1958* (Washington: History and Museums Division, HQMC (rev. 1983), and BGen Edwin H. Simmons, USMC (Ret), *The United States Marines: The First Two Hundred Years, 1775-1975* (New York: The Viking Press, 1974), pp. 255-57.

2. The Phibron would become the Mediterranean Amphibious Ready Group (MARG) once it passed from the operational control of the Second Fleet to the Sixth Fleet.

3. 32d MAU Situation Report No. 7, for the period 22-28May82, dtd 29May82 (Archives, MCHC, Washington, D.C.), hereafter 32d MAU SitRep, with number and date. See Appendix B for the task organization for each MAU deployed to Lebanon. Unless otherwise noted, all official documentation is located in the Archives, Marine Corps Historical Center, Washington Navy Yard, Washington, D.C., and the interviews in the Oral History Collection in the same place. Similarly, the Command Chronologies (ComdC) of the MAUs are held in the Archives.

4. Captain White wore two hats—one as the commodore of Phibron 4, the second as Commander, Task Force 61 (CTF 61). Similarly, Colonel Mead was two-hatted—Commanding Officer, 32d MAU, and CTF 62. The chain of command acsended from CTF 62 up through CTF 61; to Commander, Sixth Fleet; to Commander, U.S. Naval Forces, Europe (CinCUSNavEur); to Commander in Chief, U.S. Forces, Europe (CinCEur); to the Joint Chiefs of Staff in Washington.
5. Col James M. Mead, "The Lebanon Experience," *Marine Corps Gazette*, Feb83, p. 30, hereafter Mead, Lebanon.
6. 32d MAU SitRep, No. 9, dtd 12Jun82.
7. 32d MAU ComdC, Jun82.
8. 32d MAU SitRep, No. 9, dtd 12Jun82.
9. Mead, Lebanon, p. 31. This is a classified tactical situation booklet concerning noncombatant evacuation operations among other matters, and published by FMFLant.
10. LtCol Dennis R. Blankenship intvw, 12Jan83, p. 5, hereafter Blankenship intvw. The articles referred to are Col Sydney H. Batchelder, Jr., and Major David A. Quinlan, "Operation Eagle Pull," *Marine Corps Gazette*, May76, and BGen Richard E. Cary and Maj David A. Quinlan, "Frequent Wind," *Marine Corps Gazette*, Feb, Mar, and Apr76.
11. Blankenship intvw, p. 6.
12. Maj William H. Barnetson intvw, 12Jan83, p. 11, hereafter Barnetson intvw. The MSSG ran the Evacuation Control Center on the *Hermitage*, while BLT 2/8 ran the one on the *Nashville*. All similarly deployed MSSGs have been trained to conduct these operations when so ordered.
13. Blankenship intvw, p. 10
14. Mead, Lebanon, p. 31.
15. Ibid., pp. 32-33.
16. LtCol Robert B. Johnston intvw, pp. 1-2, dtd 13Jan83, hereafter Johnston intvw.
17. Ibid.
18. Sec III, 32d MAU AAR for Beirut, Lebanon Evacuation Operations, 16Aug-10Sep82, dtd 15Sep82, p. 2, hereafter 32d MAU AAR, Sep82.
19. Johnston intvw, p. 3.
20. Ibid., p. 4.
21. Ibid.
22. Ibid., p. 5.
23. Mead, Lebanon, p. 33.
24. 32d MAU SitRep, No. 19, dtd 22Aug82.
25. Mead, Lebanon, p. 33.
26. Ibid.
27. Johnston intvw, p. 13.
28. Ibid.
29. Sec II, 32d MAU AAR, Sep82, p. 8; 32d MAU ComdC, Aug82.
30. Blankenship intvw, p. 21.
31. Johnston intvw, p. 14.

32. Capt Kenneth T. McCabe intvw, dtd 14Jan83, p. 32, hereafter McCabe intvw.
33. Quoted in Mead, Lebanon, p. 36.
34. After the Italians landed and moved to positions in southern Beirut, near Galerie Semaan, they were augmented by Marine ANGLICO and liaison teams. 32d MAU ComdC, Aug82.
35. BLT 2/8 Initial AAR, encl 1 to 32d MAU AAR, Sep82, p. 4.
36. Ibid., p. 5. Plastic flags to be flown from vehicles' antennae were soon received by the MAU, but they did not hold up well. Later FMFLant authorized the wearing of American flag patches on the left shoulder of both flight suits and camouflage utilities. The wearing of the patch by members of the U.S. Multi-National Forces was for operational identification purposes only, and FMFLant considered that the patch was just one more visible means to enhance the MAU as "presence force." The flag patch also identified the Marines as Americans when conducting joint patrols with the Lebanese and especially in the event of boundary/zone disputes. Each Marine in the MAU was issued three patches to be sewn on his utilities and/or flight suits. These were to be removed upon completion of his deployment to Lebanon. CG FMFLant msg to CMC, dtd 24May83.
37. Maj Jack L. Farmer comments, 13Feb84.
38. Mead, Lebanon, p. 37.
39. Blankenship intvw, p. 23.
40. Johnston intvw, p. 16.
41. Ibid., p. 17.
42. Ibid., p. 18.
43. Ibid., p. 19.
44. Ibid.
45. U.S. Defense Attache, Beirut, msg to CMC, dtd 30Aug82.
46. Sec III, p. 1. 32d MAU AAR, Sep82.
47. Sec IV, pp. 1-3, Ibid.
48. BLT 2/8 Initial AAR, p. 4, in 32d MAU AAR, Sep82.
49. Ibid., p. 5.
50. Ibid., p. 6.
51. HMM-261 AAR, p. 2, in 32d MAU AAR, Sep82.
52. 32d MAU ComdC, Sep82.
53. Mead, Lebanon, p. 38.

Chapter 3
Footnotes

1. Upon the withdrawal of the MNF from Lebanon earlier in the month, Israeli forces returned to east Beirut and then entered west Beirut ostensibly to provide security for the PLO families remaining behind after the evacuation of the PLO guerrillas. "In reality, they were opportunistically going ahead with their desire to remove all caches from that area. They knew that substantial amounts of weapons and ammunition still existed in the area. This was fully substantiated later." Mead, Lebanon, p. 38.
2. Sec III, p. 1, 32d MAU AAR for Beirut, Lebanon Operation, 29Sep-1Nov82, dtd 12Nov82, hereafter 32d MAU AAR, Sep-Nov82.
3. 32d MAU ComdC, Sep82.
4. USCinCEur msg to CinCUSNavEur dtd 25Sep82.
5. *Long Commission Rpt*, p. 7.
6. Ibid.
7. 32d MAU AAR, Sep-Nov82.
8. *Long Commission Rpt*, p. 7.
9. As a matter of record, at no time ever during their deployment did the MAUs downgrade the threat of terrorism or fail to recognize its ever-presence in Beirut. This is apparent in a review of training schedules of the various MAUs as well as the comments made during the course of oral history interviews beginning with the 32d/22d MAU and successive MAUs concerning their deployments to Lebanon.
10. Mead, Lebanon, p. 39.
11. Blankenship intvw, p. 41. Shortly after the deployment of the 24th MAU in November, a Marine public affairs bureau was established at the MAU headquarters with a Marine public affairs officer in charge.
12. Johnston intvw, p. 26.
13. Blankenship intvw, pp. 42-43.
14. During the 1958 landing, the Marines designated this Red Beach.
15. Maj Jack L. Farmer intvw, dtd 20Dec83, pp. 1-2, hereafter Farmer intvw I.
16. Ibid.
17. Ibid., pp. 4-5.
18. Ibid.
19. 32d MAU ComdC, Sep-Nov82. It was perceived by LtCol Johnston that the landing at the port was primarily for ceremonial reasons, as the MAU wanted to land across the beach and by air at the airport. Johnston intvw, p. 5.
20. National Military Command Center msg to ComSixthFlt, dtd 30Sep82.
21. Farmer intvw I, p. 6.
22. Ibid., p. 7.
23. Although the 32d MAU's artillery was not landed, its artillerymen came ashore organized as a provisional infantry company.
24. Ibid., p. 9.
25. Johnston intvw, p. 47.

26. The description of the MAU headquarters compound and buildings were derived from the Long Commission Rpt, pp. 69-86.
27. Another club was established at the opposite end of the building for sergeants and below, and in the MSSG building, still another club was set up and entitled "The Peace Keeper's Tavern." There was also a beer hall in the BLT headquarters building.
28. Farmer intvw I, p. 12.
29. Sect II, p. 1. 32d MAU AAR, Sep-Nov82.
30. HMM-261 AAR, p. 1, in 32d MAU AAR, Sep-Nov82.
31. MSSG-32 AAR, p. 3, in 32d MAU AAR, Sep-Nov82.
32. 32d MAU ComdC, Sep82.
33. Barnetson intvw, pp. 2-3.
34. Ibid., p. 16.
35. For a fuller description of MREs, see Capt Paul Loschiavo, "Ready to Eat Meals Replace C-rats," *Marine Corps Gazette*, Jun83, pp. 33. The BLT later commented that ". . . the new MRE was well received. It was tasty but the main portion offered substantially less than the old C-rations. The absence of cans or a plausible stove was found to be a problem and consequently most Marines ate MREs cold." BLT 2/8 AAR, p. 9, in 32d MAU AAR, Sep-Nov82.
36. 32d MAU ComdC, Sep82.
37. Mead, Lebanon, p. 40.
38. Maj Christopher M. Arey intvw, dtd 17Mar83, p. 6, hereafter Arey intvw.
39. 32d MAU SitRep, No. 32, dtd 21Nov82.
40. McCabe intvw, p. 35.
41. Ibid.
42. Blankenship intvw, p. 59.
43. CMC msg to COs, 32d MAU, 24th MAU, dtd 5Nov82.

Chapter 4
Footnotes

1. Col Thomas M. Stokes, Jr., intvw, dtd 15Mar83, hereafter Stokes intvw.
2. 24th MAU SitRep No. 13, dtd 3Oct82.
3. Stokes intvw, p. 2.
4. Stokes intvw, p. 6.
5. Ibid.
6. 24th MAU SitRep No. 19, dtd 14Nov82.
7. 24th MAU ComdC, Nov82.
8. Arey intvw, p. 10.
9. Maj John A. Tempone and Capt Charles T. Botkin intvw, dtd 15Mar83, p. 43, hereafter Tempone/Botkin intvw.
10. 24th MAU ComdC, Nov82.
11. 24th MAU SitRep No. 19, dtd 14Nov82.
12. LtCol John B. Matthews intvw, dtd 16Mar83, p. 13, hereafter Matthews intvw.
13. Ibid.
14. Ibid., p. 15. As Marines use it, "arrrugah" is akin to a war cry, an expression of enthusiasm and esprit de corps. When the author was with the 22d MAU at Beirut International Airport in May 1983, he noted groups of LAF recruits jogging in the vicinity of the MAU headquarters, chanting as they ran, while cautiously looking to see whether the Marines noted and approved.
15. Ibid., p. 14.
16. Tempone/Botkin intvw, p. 26.
17. Matthews intvw, p. 37.
18. Stokes intvw, p. 18.
19. Maj David N. Buckner intvw, dtd 18Mar83, p. 21, hereafter Buckner intvw.
20. Ibid., p. 26.
21. Stokes intvw, p. 49.
22. 24th ComdC, Jan83.
23. 24th MAU SitRep No. 31, dtd 6Feb83.
24. 24th MAU ComdC, Jan83, p. 2.
25. Ibid.
26. Ibid.
27. The British were there because they were going to move into a position directly across the road from the Marines.
28. Capt Charles B. Johnson intvw, dtd 16Mar83, p. 21, hereafter Johnson intvw.
29. Ibid., p. 23.
30. Ibid., p. 25.
31. Johnson interview on station WRC, Channel 4, Washington, D.C. "5:30 Live," television program, 29Mar83.
32. Ibid.
33. Johnson intvw, p. 28.
34. Ibid., p. 30.
35. Ibid., pp. 34-36.
36. 24th MAU ComdC, Feb83.
37. 24th MAU SitRep No. 37, dtd 21Mar83.
38. Stokes intvw, p. 43.
39. Ibid.
40. Buckner intvw, p. 21.
41. Ibid., pp. 33-34.
42. Matthews intvw, p. 36.

Chapter 5
Footnotes

1. Frocking is the procedure of promoting a selected officer to the next higher grade, permitting him to wear the insignia, but not paying him the salary until

there is a vacancy for him to fill in the new grade.
2. 22d MAU SitRep No. 1, dtd 3Dec82.
3. Buckner intvw, p. 8.
4. Ibid., p. 9.
5. Ibid., p. 2.
6. 22d MAU SitRep No. 9, dtd 29Jan83.
7. Col James M. Mead, "Lebanon Revisited," *Marine Corps Gazette* Sep83, p. 67, hereafter Mead, Lebanon II.
8. 22d MAU SitRep No. 12, dtd 20Feb83.
9. Ibid.
10. To "strongback" a tent is to provide a wooden frame for rigid support.
11. Farmer intvw I, p. 16.
12. 22d MAU ComdC, Feb83.
13. Maj Jack L. Farmer intvw dtd 26May83, p. 1, hereafter Farmer intvw II.
14. Mead, Lebanon II, p. 68.
15. Farmer intvw, p. 20.
16. Farmer intvw II, p. 2.
17. 22d MAU ComdC, Feb83.
18. Mead, Lebanon II, pp. 68-69.
19. Farmer intvw II, pp. 16-17.
20. Ibid., p. 18.
21. Mead, Lebanon II, p. 69.
22. 22d MAU ComdC, Feb83.
23. Maj Jack L. Farmer comments to author, 6Mar84.
24. The reports of neither of these intelligence surveys, or ones made later by Department of Defense and National Security Agency teams, are available for publication.
25. Farmer intvw II, p. 10
26. There were indications that the PLO had drifted back into Beirut and neighboring areas of Syria.
27. Farmer intvw I, p. 22.
28. Ibid., p. 23.
29. Ibid., p. 24.
30. LtCol Donald F. Anderson intvw, dtd 25May83, p. 17, hereafter Anderson intvw.
31. 22d MAU ComdC, Mar83.
32. Mead, Lebanon II, p. 70.
33. Ibid.
34. Farmer intvw II, p. 21.
35. 22d MAU SitRep No. 17, dtd 27Mar83.
36. Farmer intvw I, p. 28.
37. Ibid.
38. 22d MAU ComdC, Mar83.
39. 22d MAU SitRep No. 19, dtd 9Apr83.
40. Farmer intvw II, p. 24.
41. Col James M. Mead intvw, dtd 23May83, p. 13, hereafter Mead intvw I.
42. BGen James Mead presentation to Marine Corps History and Museums Division, MCHC, Washington, D.C., dtd 14Sep83, p. 34, hereafter Mead Presentation.
43. Mead, Lebanon II, p. 71.
44. Farmer intvw II, p. 25.
45. A special award was established by his friends and admirers in memory of LtCol Leftwich, who was killed during his second tour in Vietnam. The Leftwich Trophy, sculpted by Iwo Jima Memorial sculptor Felix de Weldon, is awarded annually to the outstanding ground combat element captain serving in the Fleet Marine Force. In 1984 it was awarded to Major Robert K. Dobson, Jr., who won the award for services as a captain, commanding Company G, BLT 2/8, 22d MAU, in the Grenada operation and Beirut deployment.
46. 22d MAU SitRep No. 21, dtd 24Apr83.
47. Farmer intvw II, p. 26.
48. BGen James M. Mead, "Through the Looking Glass" (Unpublished ms, 1984), p. 27.
49. SecState msg to CMC, dtd 4May83.
50. 22d MAU ComdC, May83; 22d MAU SitRep No. 23, dtd 8May83.
51. Ibid.
52. Farmer intvw II, p. 28.
53. Long Commission Rpt, p. 30.
54. LtCol Arthur S. Weber memo for the Director, Plans Div, HQMC, dtd 2Jun83, p. 3.
55. Mead, Lebanon II, p. 73.
56. Ibid.
57. Author's notes.

Chapter 6
Footnotes

1. 24th MAU SitRep No. 1, dtd 25Mar83.
2. FMFLant LOI 5-83 for LanForSixFlt 2-83, dtd 4Feb83.
3. 24th MAU SitRep No. 9, dtd 22May83.
4. Col Timothy J. Geraghty intvw dtd 28May83, pp. 4-5, hereafter Geraghty intvw I.
5. 24th MAU ComdC, Jun83.
6. Long Commission Rpt, pp. 74-75.
7. 24th MAU ComdC, Jun83.
8. 24th MAU SitRep No. 13, dtd 19Jun83.
9. 24th MAU ComdC, Jun83, p. 2-9.
10. Ibid., pp. 3-4
11. Geraghty intvw I, pp. 7-8.
12. 24th MAU ComdC, Jun83, p. 2-3.
13. Ibid., p. 2.
14. 24th MAU ComdC, Aug83, p. 2-2.
15. Ibid., p. 2-3.

16. Ibid.
17. Ibid., pp. 2-4
18. Ibid.
19. Earlier recognizing that the situation in Beirut was becoming such that the time would come when they would have to return fire in self defense, Commodore France and Colonel Geraghty jointly developed an alert system for both Phibron ships and the Marines ashore, with Condition 1 being the highest state of alert and 4 the least.
20. 24th ComdC, Aug83.
21. Ibid., p. 2-8.
22. Maj Robert S. Melton intvw, dtd 19Nov83, p. 4, hereafter Melton intvw.
23. Ibid., p. 28.
24. Ibid., p. 2.
25. Ibid., p. 27.
26. 24th MAU ComdC, Sep83.
27. LtCol Harold W. Slacum intvw, dtd 21Nov83, p. 5, hereafter Slacum intvw.
28. 24th MAU SitRep No. 25, dtd 11Sep83.
29. 24th MAU ComdC, Sep83, pp. 2-3.
30. Ibid.
31. Ibid.
32. Long Commission Rpt, p. 32.
33. 24th MAU ComdC, Sep83, p. 2-6.
34. 24th MAU SitRep No. 25, dtd 11Sep83.
35. Ibid.
36. Mead Presentation, pp. 37-38.
37. Ibid., p. 39.
38. 24th MAU ComdC, Sep83.
39. 24th MAU SitRep No. 26, dtd 18Sep83.
40. Ibid.
41. Slacum intvw, pp. 9-12.
42. Col Timothy J. Geraghty intvw dtd 2Nov83, pp. 11-14, hereafter Geraghty intvw II.
43. 24th ComdC, Sep83, p. 2-7.
44. The Cobra belonged to HMM-165, the air combat element of the 31st MAU, embarked on the *Tarawa*.
45. Ibid., p. 2-9.
46. Ibid., p. 2-10.
47. 24th MAU SitRep No. 28, dtd 2Oct83.
48. 24th MAU ComdC, Oct83, pt 2, p. 2-3.
49. 24th MAU SitRep No. 30, dtd 16Oct83.
50. LtCol Laurence C. Medlin intvw dtd 20Nov83, pp. 10-11, hereafter Medlin intvw.
51. 24th MAU ComdC, Oct83, pt 2, p. 2-4. The MAU operation report for the 15th notes that there was one observed kill.
52. CTF 62 msg to CTF 61 dtd 230500Z. This message was immediately passed up the line of command by CTF 61, who amplified it with the information that the explosive was caused by a car bomb or truck carrying explosives. Commodore France also notified Washington that he had activated medevac helicopters as well as the medical teams on all Phibron 8 shipping.
53. Geraghty intvw II, pp. 18-19.
54. Ibid.
55. Ibid., p. 19.
56. Slacum intvw, pp. 18-19.
57. This is a report submitted when U.S. forces are attacked by a hostile force.
58. Geraghty intvw II, p. 20. Within 24 hours, Lieutenant Colonel Edwin C. Kelley, Jr., had arrived with his BLT 2/6 headquarters—which was now redesignated BLT 1/8—and was joined on 30 October by his Company E. 24th ComdC, Oct83, pt 3, p. 3-2.
59. Slacum intvw, p. 20.
60. Ibid., p. 22.
61. 1stLt Stephen N. Mikolaski intvw, dtd 20Nov83, p. 6, hereafter Mikolaski intvw.
62. CTF 61 msg to NMCC 230743Z. The Beirut local time of this message was 0943. For his mass casualty evacuation plan, Commodore France had previously made arrangements for casualties to be flown to the RAF hospital on Cyprus.
63. LCdr George W. Pucciarelli intvw, dtd 21Nov83, pp. 9-10, hereafter Pucciarelli intvw.
64. Ibid., pp. 11-12.
65. Ibid.
66. Mikolaski intvw, p. 18.
67. Melton intvw, p. 15.
68. CMC comments on draft ms, dtd 23Jun86, hereafter CMC comments.
69. Ibid.
70. LtCol Frank Libutti intvw, dtd 7Feb84, p. 1, hereafter Libutti intvw.
71. Eve Zibart, quoting General Kelley in "Wounded 'Semper Fi' Marine gets a General's Four Stars," *Washington Post*, 16Nov83, p. 1.
72. Libutti intvw, pp. 7-8.
73. Ibid., p. 16.
74. CMC comments.
75. General Paul X. Kelley remarks to Senate Armed Services Committee, dtd 31Oct83, p. 1, hereafter CMC Statement. See Appendix D for the complete statement.
76. Ibid., p. 3.
77. Ibid., p. 12.
78. Ibid., pp. 15-16.
79. Ibid.
80. Ibid., p. 17.
81. Ibid., pp. 17-18.

82. Ibid., pp. 18-19. Ironically, upon his return to Camp Lejeune with the 22d MAU in June 1983, Major Farmer remained for the changeover of staffs and then was transferred to Headquarters, U.S. Marine Corps, where he was assigned as head of the Casualty Section. He had not been in this new billet two weeks before the attack occurred. Major Farmer and staff set up a 24-hour telephone watch to take the thousands of calls received from anxious relatives. This condition also prevailed in the Division of Public Affairs.
83. Ibid., p. 20.
84. CMC msg to 24th MAU, dtd 9Nov83.

Chapter 7
Footnotes

1. Unless otherwise indicated the material in this section is largely derived from the *Long Commission Rpt.*
2. *Long Commission Rpt*, pp. 17-18.
3. Ibid., pp. 134-135.
4. Ibid., pp. 135-136.
5. U.S. Congress, House, Committee on Armed Services, Adequacy of U.S. Marine Corps Security in Beirut, 98th Cong., 1st sess., 1983, hereafter, *House Rpt.*
6. Ibid., pp. 1-2.
7. U.S. Congress, House, Committee on Armed Services, Adequacy of U.S. Marine Corps Security in Beirut, Summary of Findings and Conclusions, 98th Cong., 1st sess., 1983, p. 1, hereafter *House Summary*.
8. Ibid.
9. Ibid., p. 2.
10. Ibid., p. 3.
11. Ibid.

Chapter 8
Footnotes

1. 24th MAU SitRep No. 31, dtd 31Oct83.
2. CTF 62 msg to CMC, dtd 7Nov83.
3. 24th MAU ComdC, Oct83, pt 2, p. 2-7.
4. 24th MAU SitRep No. 31, dtd 31Oct83.
5. 24th MAU SitRep No. 32, dtd 6Nov83.
6. 24th MAU SitRep No. 33, dtd 13Nov83.
7. White House msg to CO, 24th MAU, dtd 10Nov83.
8. Pucciarelli intvw, p. 18.
9. 24th MAU SitRep No. 35, dtd 21NOv83.
10. 24th MAU SitRep No. 37, dtd 5Dec83.
11. CMC msg to 24th MAU, dtd 29Nov83.

Chapter 9
Footnotes

1. 22d MAU Preliminary AAR for Operation Urgent Fury; Operational Summary of Landing Force Participation, dtd 1Nov83.
2. Col James P. Faulkner intvw dtd 13Nov83, hereafter Faulkner intvw I. For the complete story of the Marine operation in Grenada, see LtCol Ronald A. Spector, *U.S. Marines in Grenada, 1983* (Washington, D.C.: HQMC, 1987).
3. CG, FMFLant msg to 22d MAU, dtd 3Nov83.
4. BGen Jim R. Joy intvw dtd 26-27May84, p. 75, hereafter Joy intvw. A MAB is usually commanded by a brigadier general.
5. Ibid., p. 76.
6. CinCEur msg to Gen Joy dtd 9Nov83.
7. 22d MAU SitRep No. 11, dtd 19Nov83.
8. CTF 62 msg to Commander, Naval Construction Battalion 1, dtd 19Nov83.
9. 22d MAU SitRep No. 11, dtd 19Nov83.
10. CG, 22d MAU msg to CGFMFLant, dtd 23Nov83.
11. 22d MAU Post-Deployment Rpt, p. 21.
12. 22d MAU SitRep No. 12, dtd 27Nov83.
13. CTF 62 msg to CGFMFLant, dtd 25Nov83.
14. LtCol Granville R. Amos intvw, dtd 22May84, pp. 6-7, hereafter Amos intvw.
15. Ibid., p. 8.
16. Maj Richard J. Gallagher intvw, dtd 22May84, pp. 6-7, hereafter Gallagher intvw.
17. 22d MAU ComdC, 1Jan-11May84, pt 3, p. 3.
18. 22d MAU SitRep No. 13, dtd 4Dec83.
19. LtCol Smith quoted in *Newsweek*, 19Dec83, p. 40.
20. 22d MAU ComdC, 5Jul-31Dec83.
21. LtCol Ray L. Smith intvw, dtd 21May84, p. 4, hereafter Smith intvw.
22. Directly across from Company G was Burj al Barajinah, dubbed "Hooterville" by the 32d MAU Marines in September 1982 and known as such by Beirut Marines ever since. On the corner of an alley opposite MAU positions was what appeared to be a restaurant with a sign over the door reading, "Cafe Daniel." As Lieutenant Colonel Smith related, however, "We were there three months before I discovered this, that actually 'Cafe Daniel' is a brand of coffee." Smith intvw, p. 9.
23. Ibid., p. 10.
24. Ibid., p. 12.

25. LtCol Earnest A. Van Huss intvw, dtd 21May84, p. 30, hereafter Van Huss intvw.
26. Ibid., p. 37.
27. Smith intvw, p. 27.
28. Ibid., p. 28.
29. Ibid., pp. 28-29.

30. 22d MAU SitRep No. 15, dtd 20Dec83. The Dragon Teeth were blocks of concrete, approximately 4'x4'x2' in size, and solid obstacles to any vehicle attempting to crash into the MAU areas.

31. Ibid.
32. Commenting on this matter, Colonel Faulkner said, "We had . . . different type [s] of threat [s] daily. And I think you have to choose which one you're going to counter today, which one you're going to take action against." Col James P. Faulkner intvw, dtd 25May84, p. 27, hereafter Faulkner intvw II.
33. 22d MAU SitRep No. 15, dtd 20Dec83.
34. Amos intvw, pp. 13-14.
35. 22d MAU SitRep No. 16, dtd 26Dec83.
36. The "vital area" was that area where the MAU, MSSG, and the BLT headquarters had been located prior to the bombing. After the bombing and until it re-embarked in February, the MSSG elements which had not gone back on board ships earlier, remained in place. The BLT and the MAU remained in the vital area until their new dug-in bunkers were ready for occupation in the southwest portion of the airport. Throughout the post-bombing period, the vital area was protected by significant defenses—the tank ditch, Dragon Teeth, and wire obstacles. A rifle company defended the area, augmented at night by an engineer platoon. Capt Christopher J. Guenther intvw, dtd 22May84, pp. 29-30, hereafter Guenther intvw.
37. 22d MAU SitRep No. 16, dtd 26Dec83.
38. LtCol Ronald R. Rice intvw, dtd 24May84, p. 5, hereafter Rice intvw.
39. Ibid., p. 8.
40. 22d MAU ComdC, 5Jul-31Dec83, pt 3, p. 11.
41. 22d MAU SitRep No. 17, dtd 2Jan84.
42. Rice intvw, p. 9.
43. 1stLt Michael L. Ettore intvw, dtd 22May84, p. 2, hereafter Ettore intvw.
44. Ibid., pp. 4-5.
45. Ibid., p. 9.
46. Ibid., p. 11.
47. Ibid., pp. 11-12.
48. Joy intvw, p. 59.
49. 22d MAU SitRep No. 19, dtd 16Jan84.
50. 22d MAU ComdC, 1Jan-11May84, pt 3, p. 2.
51. 22d MAU SitRep No. 19, dtd 15Jan84.
52. 22d MAU ComdC, 1Jan-11May84, loc cit.
53. Ibid., p. 6.
54. 22d MAU SitRep No. 21, dtd 29Jan84.
55. 22d MAU SitRep No. 22, dtd 12Feb84.
56. 22d MAU SitRep No. 22, dtd 5Feb84. Because of a numbering error the MAU issued two SitReps numbered "22": one on 5Feb84 and the second on 12Feb84.
57. Ibid. EUCOM engineers and logisticians had earlier recommended that the bunkers being built in Phase II of the MAU construction plan were to be hardened to provide adequate protection from delayed fuze-detonated ordnance, which had not yet been used against the Marines. Accordingly, the materials to accomplish this were requisitioned through channels. When JCS Chairman General Vessey visited the MAU headquarters and toured the Marine positions on 8 January, he was not convinced that Phase III needed to be fulfilled. Although he made no decision one way or another, completion of Phase III construction was overtaken by events. Nonetheless, a lot of material necessary for Phase III construction had already been delivered and more was in the pipeline. According to General Joy, ". . . if we had completed the whole construction plan of Phase III, we would have spent about $7.5 million for construction. As it turned out, we spent about $4.5 million." Joy intvw, p. 56.
58. Smith intvw, pp. 35-36.
59. Ibid.
60. 22d MAU SitRep No. 22, dtd 12Feb84.
61. Ibid.
62. 22d MAU ComdC, 1Jan-11May84, dtd 11May84, pt 1, p. 2.
63. 22d MAU SitRep No. 22, dtd 12Feb84.
64. Van Huss intvw, p. 47.
65. Ibid., p. 48.
66. Joy intvw, pp. 66-67. The headquarters of the British contingent was located across the Old Sidon Road from the Marine outpost at Lebanon University. When the Americans were withdrawn from here and put on board ships in November 1983, the British had found themselves rather isolated from the rest of the MNF units.
67. Van Huss intvw, pp. 48-49.
68. BGen Joy telecon with author, 31Jul84.
69. Joy intvw, p. 62.
70. 22d MAU SitRep No. 23, dtd 19Feb84.
71. LtCol Douglas M. Davidson intvw, dtd 24May84. p. 27, hereafter Davidson intvw.
72. Ibid., p. 28.

73. MSSG 22 Post-Deployment Rpt, pt 1, p. 9, dtd 6Apr84, enclosure (3) to 22d MAU Post-Deployment Rpt. LtCol Davidson was not an advocate of seabasing, particularly for prolonged operations. "It's good for short duration exercises where you don't have time to put everything ashore anyway." Davidson intvw, p. 14.

74. Maj Stephen D. Anderson intvw, dtd 21May84, p. 18.

75. Amos intvw, pp. 22-23.
76. Van Huss intvw, p. 56.
77. Ibid., pp. 56-57.
78. 22d MAU SitRep No. 25, dtd 5Mar84.
79. Van Huss intvw, pp. 61-62.
80. Ibid.
81. 22d MAU ComdC, 1Jan-11May84, dtd 11May84, pt 3, p. 13
82. The material in this section about deactivation of JTFL was derived from Joy intvw, pp. 72-73.

Appendix A

Chronology: Marines in Lebanon, 1982-1984

1982

25 August	Roughly 800 Marines of the 32d Marine Amphibious Unit (MAU), commanded by Colonel James M. Mead, landed in Beirut as part of a multinational peacekeeping force to oversee evacuation of Palestine Liberation Organization (PLO) guerrillas. The force also includes 400 French and 800 Italian soldiers.
10 September	Evacuation of PLO complete; 32d MAU was ordered out of Beirut by the President of the United States.
26 September	Preparation for redeployment of the 32d MAU to Beirut got under way, in the wake of the assassination of Lebanese President-elect Bashir Gemayel, an Israeli push into Moslem West Beirut, and the massacre of Palestinians at the Sabra and Shatila refugee camps. Marines and sailors of the 32d MAU received the Navy Unit Commendation for their part in the PLO evacuation, in ceremonies on board the USS *Guam* (LPH-9), 60 miles off the coast of Lebanon.
29 September	The 32d MAU returned to Beirut, to join 2,200 French and Italian troops already in place.
30 September	Marines suffer first casualties (one killed in action, three wounded in action) while clearing unexploded ordnance from the vicinity of Beirut International Airport.
30 October	The 32d MAU was relieved by the 24th MAU, commanded by Colonel Thomas M. Stokes, Jr.
4 November	The 24th MAU extends its presence in Beirut to the eastern (Christian) sector, patrolling the "Green Line" that divides the city into sectarian parts.
3 December	24th MAU artillery was moved ashore (Battery of six 155mm howitzers).
10 December	24th MAU armor was moved ashore (Platoon of five M60A7 tanks).
13 December	Marines commence training of Lebanese Armed Forces. About 75 Lebanese soldiers underwent 21 days' training in basic infantry skills and helicopter assaults.

1983

29 January	Emergency communication network established between American and Israeli forces as tensions mounted between adjacent ground units.
2 February	Captain Charles B. Johnson confronted three Israeli tanks as they attempted to pass through his company check point.

15 February The 32d MAU, redesignated the 22d MAU and still commanded by Colonel Mead, returned to Lebanon to relieve the 24th MAU.

21 February Marines commence four days of relief operations in the town of Quartaba during Lebanon's worst blizzard in memory. With Syrian acquiescence, Marine helicopters also flew into Syrian-held territory in Lebanon's central mountains to rescue victims of frostbite and exposure.

16 March Five Marines were wounded in action in first direct attack on American peacekeeping troops. An Islamic fundamentalist group claims responsibility.

17 March The 24th MAU received the Navy Unit Commendation for Lebanon service from October 1982 to February 1983.

18 April A large car bomb explodes at the U.S. Embassy in Beirut, causing massive structural damage and killing 61, including 17 Americans. More than 100 were injured. Islamic fundamentalists again claim responsibility.

5 May Marine helicopter with six aboard, including Colonel Mead, is hit by ground fire as it investigates artillery duels between Druze and Christian gunners.

17 May Lebanon-Israeli withdrawal agreement is signed.

30 May The 24th MAU, commanded by Colonel Timothy J. Geraghty, relieves the 22d MAU.

25 June Marines conduct first combined patrols with Lebanese Army troops.

27 June The 22d MAU received the Navy Unit Commendation for Lebanon service from 15 February to 30 June 1983.

22 July Two Marines and one sailor wounded in action by shell fragments during shelling of Beirut International Airport, part of a general pattern of increasing indirect fire against the Lebanese Army, the airport, and the multinational force.

10 August About 27 artillery and mortar rounds were fired by Druze militia from the high ground east of Beirut into Beirut International Airport, resulting in one Marine wounded in action. Rockets also hit the Defense Ministry and the Presidential Palace. Three Cabinet ministers were kidnapped by the Druze.

11 August Eight more rocket/artillery rounds fired into Beirut International Airport. No casualties.

28 August A combat outpost manned by 30 Marines and Lebanese Army troops east of Beirut International Airport came under fire from semiautomatic weapons and rocket-propelled grenades. Marines return fire for the first time, with rifles and M-60 machine guns. No friendly casualties, after a 90-minute firefight.

29 August A heavy rocket, mortar, and artillery attack on 24th MAU positions on the eastern side of Beirut International Airport resulted in two Marines killed in action and 14 Marines wounded in action. Marines retaliate with 155mm artillery.

30 August French and Italian command posts hit by mortar fire. One French soldier killed, five Italians wounded.

31 August	Department of Defense authorized hostile fire pay of $65 per month for Marines and sailors of the 24th MAU serving in Lebanon.
31 August	Marines retaliate with 155mm artillery after Moslem shelling of U.S. Embassy residence.
1 September	Joint Chiefs of Staff directed deployment of Amphibious Ready Group Alpha with the 31st MAU embarked, from the Western Pacific to the Mediterranean, in the vicinity of Lebanon.
4 September	Israeli forces withdrew to positions on the Awwali River, creating a void to be filled by factional hostilities among the Lebanese.
6 September	Rocket attack on Beirut International Airport from Druze positions in Shouf mountains resulted in two Marines killed, two Marines wounded. Total since 28 August: four KIA, 28 WIA.
8 September	Frigate USS *Bowen* (FF-1079) fired 5-inch guns in first American use of naval gunfire support, silencing a Druze militia battery that had shelled Beirut International Airport. Marines also responded with 155mm artillery fire.
10 September	Battleship USS *New Jersey* (BB-62) was alerted for deployment to the Eastern Mediterranean.
12 September	31st MAU arrived off Lebanon, assumed standby role.
16 September	Destroyer USS *John Rodgers* (DD-983) and frigate USS *Bowen* responded with 5-inch gunfire into Syrian-controlled parts of Lebanon, after continued shelling near the residence of the U.S. ambassador.
19 September	USS *John Rodgers* and USS *Virginia* (CGN-38) fire 338 five-inch rounds to help Lebanese Army troops retain hold on strategic Shouf Mountains village of Suq al Gharb. American role shifted from "presence" to direct support of Lebanese Armed Forces, in perception of rebel factions.
20 September	Residence of U.S. ambassador was shelled; USS *John Rodgers* and USS *Virginia* respond.
21 September	USS *John Rodgers* and USS *Arthur Radford* (DD-968) responded to shelling of Marines at Beirut International Airport.
23 September	Indirect fire attack on Marine positions countered by 155mm artillery fire and five-inch gunfire from USS *Virginia*.
24 September	USS *New Jersey* arrived off Lebanese coast following high-speed transit from duty off Central America.
26 September	Cease-fire went into effect at 0600. Announced by Saudi Arabian and Syrian officials in Damascus, supported by Druze. Talks begin on formation of new coalition government for Lebanon. Marine casualties to date: five killed, 49 wounded.
1 October	31st MAU departed Mediterranean for Indian Ocean, in response to threatened crisis near Strait of Hormuz.
5 October	Two Marine helicopters hit by ground fire.
8 October	Two Marines wounded by sniper fire.
13 October	One Marine wounded by grenade fragments.
14 October	One Marine killed, three wounded by sniper fire. Marine sharpshooters responded, setting off three-hour fire-fight. Ceasefire of 26 September allegedly still in place.
15 October	Marine sharpshooters kill four snipers.

16 October	One Marine killed, five wounded by sniper fire.
19 October	Four Marines wounded as attempt to ambush Marine convoy with car bomb was thwarted.
23 October	Suicide truck loaded with equivalent of 12,000 pounds of explosives destroyed headquarters building of BLT 1/8 at Beirut International Airport. Almost simultaneous suicide attack destroyed building occupied by French paratroopers. U.S. casualties: 241 killed, 70 wounded. French casualties: 58 killed. Marine replacement airlifts, via 13 C-141 aircraft, begin the same day.
25 October	Commandant of the Marine Corps General Paul X. Kelley visited wounded in West German hospital and flies on to Lebanon to inspect scene of suicide attack.
4 November	Department of Defense established commission headed by Admiral Robert L. G. Long, USN (Ret.), to investigate 23 October suicide attack at Beirut International Airport. Suicide driver blows up Israeli headquarters in Tyre, killing 29 soldiers and 32 prisoners.
7 November	Brigadier General Jim R. Joy, USMC, arrived in Beirut to assume command of Marine operations in Lebanon.
19 November	The 24th MAU was relieved by the 22d MAU, which had participated in the 25 October-2 November Grenada intervention en route to the Mediterranean. Brigadier General Joy was overall commander of Lebanon operations for the Marines.
22 November	Defense Secretary Caspar Weinberger stated that the 23 October suicide attack on the Marines was carried out by Iranians with the "sponsorship, knowledge, and authority of the Syrian government."
4 December	Marines at Beirut International Airport came under heavy fire from gun positions in Syrian-held territory. Marine casualties: eight killed, two wounded. Naval gunfire missions fired in retaliation. Earlier in the day, a 28-plane raid was conducted on Syrian antiaircraft positions in the mountains east of Beirut, in retaliation for Syrian fire directed at American aerial reconnaissance missions. Two U.S. aircraft are downed, in this first combat mission over Lebanon.
15 December	The battleship USS *New Jersey* delivered 16-inch gunfire on antiaircraft positions in the Syrian-occupied mountains southeast of Beirut, as the Syrians continue to fire at U.S. reconnaissance flights over the area. This was the USS *New Jersey*'s first action off Lebanon.
28 December	The Long Commission released an unclassified 140-page report on the 23 October suicide attack.

1984

8 January	A Marine is killed by unidentified assailants as he exits a helicopter at a landing zone on the edge of downtown Beirut. The helicopter flew to safety, after returning fire with its machine guns.
13 January	Marines in the Beirut International Airport area fought a 30-minute battle with gunmen firing from a building east of their perimeter.
15 January	Druze gunners closed Beirut International Airport for three hours with intense 23mm fire on Marine positions east and southeast of

the airport. U.S. forces responded with small arms fire, mortars, rockets, tank fire, and naval gunfire from the battleship USS *New Jersey* and destroyer USS *Tattnall*. No U.S. casualties.

2 February	Heavy fighting erupted in the suburbs of Beirut, between the Lebanese Army and Shiite militiamen.
3 February	Shiite leadership called for resignation of Moslem cabinet members and urges Moslems in the Lebanese Army to disregard the orders of their leaders. Prime Minister Wazzan and the Lebanese cabinet resigned, to clear way for formation of new coalition government.
6 February	Druze and Moslem militiamen seized much of Beirut in street fighting and demanded resignation of Gemayel.
7 February	President Reagan announced decision to redeploy Marines from Beirut International Airport to ships offshore, leaving a residual force behind to protect the U.S. Embassy and other American interests. Increased reliance on air strikes and naval gunfire support indicated.
8 February	USS *New Jersey* bombarded Druze and Syrian gun positions as part of the heaviest naval gunfire support since the arrival of the Marines in 1982.
10-11 February	American civilians and other foreign nationals were evacuated from Beirut by helicopter.
21 February	Marines began their redeployment to ships of the Sixth Fleet offshore. About 150 Marines departed in the first increment.
26 February	Redeployment of the 22d MAU to offshore ships completed.

Appendix B
Marine Command and Staff List

32D MARINE AMPHIBIOUS UNIT

16 August-10 September; 25 September-1 November 1982

32D MARINE AMPHIBIOUS UNIT HEADQUARTERS

CO . . . . . Col James M. Mead
ExO . . . . . LtCol Charles R. Smith, Jr.
S-1 . . . . . 1stLt Michael H. Burnett
S-2 . . . . . CWO-2 Joe L. Winbush
S-3 . . . . . Maj Dennis R. Blankenship
S-4 . . . . . Maj Reuben B. Payne III

Detachment, Photographic Imagery Interpretation Unit, Marine Wing Headquarters Squadron 2, 2d Marine Aircraft Wing
Detachment, Marine Air Base Squadron 32, Marine Aircraft Group 32, 2d Marine Aircraft Wing
Subteam (-), Counterintelligence Team, Headquarters Battalion, 2d Marine Division
Detachment, Public Affairs Office, 2d Marine Division
Detachment, National Security Agency/Cryptological Security Service, Defense Systems Security Team
Detachment, Air and Naval Gunfire Liason Company, 2d Force Service Support Group

Battalion Landing Team (BLT) 2/8

CO . . . . . LtCol Robert B. Johnston

2d Battalion, 8th Marines
Battery H (Reinforced), 3d Battalion, 10th Marines
Detachment, Headquarters Battery, 3d Battalion, 10th Marines (Ships Fire Control Party)
2d Platoon, Company A, 2d Reconnaissance Battalion
2d Platoon (Reinforced), Company B, 2d Tank Battalion
Detachment, Headquarters and Service Company, 2d Tank Battalion
1st Platoon (Reinforced), Company A, 2d Assault Amphibian Battalion
Detachment, Headquarters and Service Company, 2d Assault Amphibian Battalion
2d Platoon (Reinforced), Company C, 2d Combat Engineer Battallion
Detachment, Headquarters and Service Company, 2d Combat Engineer Battalion
Detachment, Engineer Support Company, 2d Combat Engineer Battalion
3d Section, 3d Platoon, Antitank Company (TOW), 2d Tank Battalion
Detachment, Headquarters Company, 8th Marines (Multi-Channel Radio Team)

Marine Medium Helicopter Squadron 261 (-) (Reinforced)

CO . . . . . LtCol Graydon F. Geske

Marine Medium Helicopter Squadron 261
Detachment, Marine Light Helicopter Squadron 167
Detachment, Marine Attack Helicopter Squadron 269
Detachment, Marine Heavy Helicopter Squadron 362
Detachment, Headquarters and Maintenance Squadron 26
Detachment, Marine Air Base Squadron 26
Detachment, Marine Wing Service Group 27
Detachment, Headquarters and Maintenance Squadron 29
Detachment, Marine Air Base Squadron 29

Marine Amphibious Unit Service Support Group 32

CO . . . . Maj William H. Barnetson

1st Platoon (-) (Reinforced), Company A, 2d Landing Support Battalion, 2d Force Service Support Group
Detachment, Headquarters and Service Company, 2d Landing Support Battalion, 2d Force Service Support Group
Detachment, 2d Maintenance Battalion, 2d Force Service Support Group
Detachment, 2d Supply Company, 2d Force Service Support Group
Detachment, 9th Engineer Support Battalion, 2d Force Service Support Group
Detachment, 2d Medical Battalion, 2d Force Service Support Group
Detachment, 22d Dental Company, 2d Force Service Support Group
Detachment, Military Police Company, Headquarters Battalion, 2d Marine Division
Detachment, Headquarters and Service Battalion, 2d Force Service Support Group
Detachment, 8th Motor Transport Battalion, 2d Force Service Support Group

24TH MARINE AMPHIBIOUS UNIT

1 November 1982-15 February 1983

24TH MARINE AMPHIBIOUS UNIT HEADQUARTERS

CO . . . . Col Thomas M. Stokes, Jr.
ExO . . . . LtCol George T. Schmidt
S-1 . . . . 1stLt Michael K. Ritchie
S-2 . . . . 1stLt Joseph F. Ciano, Jr.
S-3 . . . . Maj John A. Tempone
S-4 . . . . Maj Frederick J. Moon

Subteam, Imagery Interpretation Unit, Fleet Marine Force, Atlantic
Subteam, Imagery Interpretation Unit, 2d Marine Division
Detachment, Counterintelligence Team, Fleet Marine Force, Atlantic
Subteam, Interrogator/Translator Team, Headquarters Battalion, 2d Marine Division
Detachment, Marine Wing Headquarters Squadron 2, 2d Marine Aircraft Wing (Photo)
Section, Sensor Control and Management Platoon, Headquarters Battalion, 2d Marine Division
Detachment, Public Affairs Division, Headquarters Marine Corps
Detachment, 2d Radio Battalion, 2d Marine Division
Detachment, Air and Naval Gunfire Liaison Company, 2d Force Service Support Group
Detachment, Communication Company, Headquarters and Service Battalion, 2d Force Service Support Group

Battalion Landing Team 3/8

CO . . . . LtCol John B. Matthews

3d Battalion, 8th Marines
Battery G, 3d Battalion, 10th Marines
3d Platoon, Company A, 2d Reconnaissance Battalion
1st Platoon, Company A, 2d Tank Battalion
2d Platoon, Company B, 2d Assault Amphibian Battalion
3d Platoon, Company C, 2d Combat Engineer Battalion
2d Section, 2d Platoon, Antitank Company (TOW), 2d Tank Battalion
Detachment, Headquarters Company, 8th Marines

Marine Medium Helicopter Squadron 263 (-) (Reinforced)

CO . . . . LtCol William G. Barnes

Marine Medium Helicopter Squadron 263
- Detachment, Marine Light Helicopter Squadron 167
- Detachment, Marine Attack Helicopter Squadron 269
- Detachment, Marine Heavy Helicopter Squadron 461
- Detachment, Marine Air Base Squadron 26
- Detachment, Marine Air Base Squadron 29
- Detachment, Headquarters and Maintenance Squadron 26
- Detachment, Headquarters and Maintenance Squadron 29
- Detachment, Marine Wing Service Group 27

Marine Amphibious Unit Service Support Group 24

CO . . . . Maj David N. Buckner

Headquarters, Marine Amphibious Unit Service Support Group 24
Detachment, 2d Maintenance Battalion, 2d Force Service Support Group
Detachment, 2d Supply Battalion, 2d Force Service Support Group
Detachment, 8th Engineer Support Battalion, 2d Force Service Support Group
Detachment, 8th Motor Transport Battalion, 2d Force Service Support Group
Detachment, 2d Dental Company, 2d Force Service Support Group
Detachment, Explosive Ordnance Demolition Team, 2d Support Battalion, 2d Force Service Support Group
Detachment, Navy Environmental Preventive Medicine Unit 2, Norfolk, Virginia
Detachment, Navy Environmental Preventive Medicine Unit 5, San Diego, California
Detachment Company B, 2d Combat Engineer Battalion, 2d Marine Division
Detachment, Utilities, 8th Engineer Support Battalion, 2d Force Service Support Group

22D MARINE AMPHIBIOUS UNIT

15 February-29 May 1983

22D MARINE AMPHIBIOUS UNIT HEADQUARTERS

CO . . . . Col James M. Mead
XO . . . . LtCol Ronald R. Rice
S-1 . . . . 1st Thomas F. Amsler
S-2 . . . . CWO-3 Joe L. Winbush
S-3 . . . . Maj Jack L. Farmer
S-4 . . . . Maj Reuben B. Payne III

Detachment, Photographic Imagery Interpretation Unit, Marine Wing Headquarters Squadron 2, 2d Marine Aircraft Wing
Subteam, Interrogator/Translator Team, Headquarters Battalion, 2d Marine Division
Detachment, Marine Air Base Squadron 32, Marine Aircraft Group 32, 2d Marine Aircraft Wing (Photo)
Detachment, Sensor Control and Management Platoon, Headquarters Battalion, 2d Marine Division

Battalion Landing Team 2/6

CO . . . . LtCol Donald F. Anderson
2d Battalion, 6th Marines
Battery I (Reinforced), 3d Battalion, 10th Marines
Detachment, Headquarters Battery, 3d Battalion, 10th Marines (Ships Fire Control Party)
3d Platoon, Company B, 2d Reconnaissance Battalion
1st Platoon (Reinforced), Company D, 2d Tank Battalion
Detachment, Headquarters and Service Company, 2d Tank Battalion
2d Platoon (Reinforced), Company A, 2d Assault Amphibian Battalion
Detachment, Headquarters and Service Company, 2d Assault Amphibian Battalion
1st Platoon (Reinforced), Company A, 2d Combat Engineer Battalion
Detachment, Headquarters and Service Company, 2d Combat Engineer Battalion
Detachment, Engineer Support Company, 2d Combat Engineer Battalion
2d Section, 1st Platoon, Antitank Company (TOW), 2d Tank Battalion
Detachment, Headquarters Company, 8th Marines (Multichannel Radio Team)

Marine Medium Helicopter Squadron 264 (-) (Reinforced)

CO . . . . LtCol Richard J. Kalata
Marine Medium Helicopter Squadron 264
Detachment, Marine Light Helicopter Squadron 167
Detachment, Marine Attack Helicopter Squadron 269
Detachment, Marine Heavy Helicopter Squadron 362
Detachment, Headquarters and Maintenance Squadron 26
Detachment, Marine Air Base Squadron 26
Detachment, Headquarters and Maintenance Squadron 29
Detachment, Marine Air Base Squadron 29

Marine Amphibious Unit Service Support Group 22

CO . . . . Maj Albert E. Shively
Headquarters, Marine Amphibious Unit Service Support Group 22
1st Platoon (-) (Reinforced), Company A, Landing Support Battalion, 2d Force Service Support Group
Detachment, Headquarters and Service Company, 2d Landing Support Battalion, 2d Force Service Support Group
Detachment, 2d Maintenance Battalion, 2d Force Service Support Group
Detachment, 2d Supply Battalion, 2d Force Service Support Group
Detachment, 8th Engineer Support Battalion, 2d Force Service Support Group
Detachment, 2d Medical Battalion, 2d Force Service Support Group
Detachment, 2d Dental Company, 2d Force Service Support Group
Detachment, Military Police Company, Headquarters Battalion, 2d Marine Division
Detachment, Headquarters and Service Battalion, 2d Force Service Support Group

24TH MARINE AMPHIBIOUS UNIT

30 May-18 November 1984

24TH MARINE AMPHIBIOUS UNIT HEADQUARTERS

CO . . . . Col Timothy J. Geraghty
ExO . . . . LtCol William A. Beebe II (to 18 July)
LtCol Harold W. Slacum (from 18 July)
S-1 . . . . 1stLt Charles F. Davis III
S-2 . . . . Capt Kevin J. McCarthy
S-3 . . . . Maj George S. Converse
S-4 . . . . Maj Robert S. Melton

Detachment, Joint Public Affairs Bureau, Headquarters Marine Corps
Detachment, 2d Air and Naval Gunfire Liasison Company, 2d Force Service Support Group
Detachment, Ashore Mobile Communications Center
Detachment, 2d Radio Battalion
Detachment, Interrogator/Translator Team
Detachment, Field Artillery School Target Acquisitions Battery

Battalion Landing Team 1/8

CO . . . . LtCol Howard L. Gerlach (to 23 October)
LtCol Edwin C. Kelley, Jr. (From 24 October)

1st Battalion, 8th Marines
1st Platoon, Company C, 2d Tank Battalion
1st Platoon, Company C, 2d Reconnaissance Battalion
3d Platoon, Company B, 2d Assault Amphibian Battalion
1st Platoon, Company C, 2d Combat Engineer Battalion
1st Section, Antitank Company (TOW), 2d Tank Battalion
Target Acquisition Battery, 10th Marines
Battery C, 1st Battalion, 10th Marines

Marine Medium Helicopter Squadron 162 (-) (Reinforced)

CO . . . . LtCol Laurence R. Medlin

Marine Medium Helicopter Squadron 162
- Detachment, Marine Light Helicopter Squadron 167
- Detachment, Marine Attack Helicopter Squadron 269
- Detachment, Marine Heavy Helicopter Squadron 464
- Detachment, Marine Air Base Squadron 29
- Detachment, Headquarters and Maintenance Squadron 26
- Detachment, Headquarters and Maintenance Squadron 29
- Detachment, Marine Wing Support Group 27

Marine Amphibious Unit Service Support Group 24

CO . . . . Maj Douglas C. Redlich

Detachment, Headquarters and Service Battalion, 2d Force Service Support Group

Detachment, 2d Maintenance Battalion, 2d Force Service Support Group
Detachment, 2d Landing Support Battalion, 2d Force Service Support Group
Detachment, 8th Engineer Support Battalion, 2d Force Service Support Group
Detachment, 8th Motor Transport Battalion, 2d Force Service Support Group
Detachment, 2d Medical Battalion, 2d Force Service Support Group
Detachment, 2d Dental Battalion, 2d Force Service Support Group
Detachment, 2d Supply Battalion, 2d Force Service Support Group
Detachment, 8th Communications Battalion, 2d Force Service Support Group
Detachment, Headquarters Battalion, 2d Marine Division
Detachment, Marine Wing Support Group 27, 2d Marine Aircraft Wing
Detachment, Communications Company, Headquarters and Service Battalion, 2d Force Service Support Group
Detachment, Postal Section, Service Company, Headquarters and Service Battalion, 2d Force Service Support Group
Detachment, Disbursing Section, Service Company, Headquarters and Service Battalion, 2d Force Service Support Group
Detachment, Beach and Port Company, 2d Landing Support Battalion, 2d Force Service Support Group
Detachment, Ammunition Company, 2d Supply Battalion, 2d Force Service Support Group
Detachment, Explosive Ordnance Disposal Team, 2d Support Battalion, 2d Force Service Support Group
Detachment, Navy Preventive Environmental Medicine Unit, Team 2, Norfolk, Virginia

22D MARINE AMPHIBIOUS UNIT

17 November 1983-9 April 1984

22D MARINE AMPHIBIOUS UNIT HEADQUARTERS

CG............................BGen Jim R. Joy (17 November-19 February 84)
CO..............................Col James P. Faulkner (from 19 February 84)
C/S......................Col James P. Faulkner (17 November 83-19 February 84)
ExO............................LtCol Ronald R. Rice (From 20 February 84)
S-1/G-1...1stLt Kenneth R. Bergman
G-2......................LtCol Forrest L. Lucy (17 November 83-19 February 84)
S-2...............................Capt Paul M. Jungel (from 20 February 84)
G-3................LtCol Edmund J. Connelly, Jr. (November 83-19 February 84)
S-3..........................LtCol Ernest A. Van Huss (from 20 February 84)
G-4..................LtCol Charles S. Rinehart (17 November 83-19 February 84)
S-4...............................Maj Albert J. Martin (from 20 February 84)

Headquarters, 22d Marine Amphibious Unit
Detachment, Photographic Imagery Interpretation Unit, Marine Wing Headquarters Squadron 2, 2d Marine Aircraft Wing
Subteam, Interrogator/Translator Team, Headquarters Battalion, 2d Marine Division

Battalion Landing Team 2/8

CO..LtCol Ray L. Smith
2d Battalion, 8th Marines
Battery H (Reinforced), 3d Battalion, 10th Marines
Detachment, Headquarters Battery, 3d Battalion, 10th Marines
3d Platoon, Company A, 2d Tank Battalion

4th Platoon (Reinforced), Company A, 2d Assault Amphibian Battalion
Detachment, Headquarters and Service Company, 2d Assault Amphibian Battalion
2d Platoon, Company C, 2d Combat Engineer Battalion
1st Section, 1st Platoon, Antitank Company (TOW), 2d Tank Battalion
1st Platoon, Company A, 2d Reconnaissance Battalion

Marine Medium Helicopter Squadron 261 (-) (Reinforced)

CO..LtCol Granville R. Amos
Marine Medium Helicopter Squadron 261
Detachment, Marine Light Helicopter Squadron 167
Detachment, Marine Heavy Helicopter Squadron 362
Detachment, Marine Headquarters and Maintenance Squadron 26
Detachment, Marine Air Base Squadron 26
Detachment, Marine Wing Service Group 27

Marine Amphibious Unit Service Support Group 22

CO...LtCol Douglas M. Davidson
Headquarters, Marine Amphibious Unit Service Support Group 22
Detachment, 8th Engineer Support Battalion, 2d Force Service Support Group
Detachment, 2d Landing Support Battalion, 2d Force Service Support Group
Detachment, 2d Maintenance Battalion, 2d Force Service Support Group
Detachment, 2d Medical Battalion, 2d Force Service Support Group
Detachment, 2d Dental Battalion, 2d Force Service Support Group
Detachment, 2d Supply Battalion, 2d Force Service Support Group
Detachment, Military Police Company, Headquarters Battalion, 2d Marine Division
Detachment, Explosive Ordnance Disposal Team, 2d Support Battalion, 2d Force Service Support Group

Appendix C
Foreign Multi-National Force Units

British Forces, Lebanon
(8Feb83-8Feb84)

Units

C Squadron, Queen's Dragoon Guards (8Feb-8Aug83)
A Squadron, Queen's Dragoon Guards (8Aug-7Dec83)
A Squadron, 16/5 Lancers (7Dec83-8Feb84)

Commanders

LtCol John C. Cochrane, Royal Irish Dragoons, 1Feb-5Aug83
LtCol David L. Roberts, Parachute Regiment, 5Aug-15Sep83
LtCol John deP. Ferguson, Queen's Dragoon Guards, 16Sep83-3Jan84
LtCol Peter E. Woolley, Prince of Wales' Own, 3Jan-8Feb84

French Forces*
(24Aug82-31Mar84)

Units and Commanders

11th Airborne Division, BGen Jacques Granger (24Aug-14Sep82)
9th Marine Infantry Division
2d Foreign Legion Airborne Battalion
3d Marine Airborne Battalion
9th Headquarters Support Battalion
17th Airborne Engineers Battalion

11th Airborne Division, BGen Jacques Granger (Sep82-Jan83)
8th Marine Airborne Battalion
1st Airborne Hussars Battalion
2d Marine Infantry Battalion
17th Airborne Engineers Battalion
1st Headquarters Support Battalion

9th Marine Infantry Division, BGen Michel Datin (Jan-May83)
2d Marine Infantry Battalion
3d Marine Infantry Battalion
11th Marine Artillery Battalion
Marine Armored Infantry Battalion

31st Brigade, BGen Jean-Claude Coulon (May-Sep83)
21st Marine Infantry Battalion
  Engineers Company, 21st Marine Infantry Battalion
2d Foreign Legion Infantry Battalion
1st Foreign Legion Cavalry Battalion
17th Airborne Engineers Battalion

11th Airborne Division, BGen Francois Cann (Sep83-Jan84)
3d Marine Airborne Battalion
6th Airborne Infantry Battalion
6th Airborne Battalion (company)
1st Airborne Infantry Battalion (company)
9th Airborne Infantry Battalion (company)
1st Airborne Hussars Battalion (platoon)
17th Airborne Engineers Battalion (company)
12th Field Artillery Battalion (battery)
7th Headquarters Airborne Support Battalion (support detachment)

9th Marine Infantry Division, BGen Datin (Feb-Mar84)
9th Headquarters Support Battalion (detachment)
501st Tank Battalion (platoon)
Marine Armored Infantry Battalion (platoon)
Gendarmerie (MPs) (platoon)
2d Marine Infantry Battalion (2 companies)
12th Field Artillery Battalion (battery) (later replaced by 68th Field Artillery Battalion)
59th Engineer Company
41st Transmission Battalion (company)

Italian Forces*
(Aug82-Jan84)

Commander: BGen Franco Angioni (Aug82-Jan84)

Units
2d Bersaglieri (Mechanized) Battalion "Governolo" (22Aug-12Sep82)
10th Bersaglieri Battalion "Bezzecca" (Feb-Jun83)
5th Airborne Battalion "El Alamein" (Feb83-Jan84)
"San Marco" Naval Infantry Battalion (27Sep82-Jan84)
"Folgore" Airborne Brigade (27Sep82-Jan84)
  1st Carabinieri Airborne Battalion "Tuscania" (two companies)
  9th Airborne Assault Battalion "Col Moschin" (one company)
Field hospital
Logistics unit (unnamed)
67th Mechanized Battalion "Montelungo" (Jun-Oct83) (four companies)
3d Bersaglieri Battalion "Cernaia" (Oct83-Jan84)

**In many cases, units identified may be components/detachments/headquarters staffs, and not the entire organization.*

Appendix D

Remarks by the Commandant of the Marine Corps Senate Armed Services Committee, 31 October 1983

On 23 October 1983, two suicidal drivers, representing interests which are totally hostile to the United States of America and the Republic of France, conducted unprecedented and massive terrorist attacks—not against American Marines, sailors, and soldiers and French airborne troops—but against the free world.

While all Americans and Frenchmen are feeling the strong emotions resulting from this act, and while I am deeply saddened by the reason for my presence before this Committee, I am relieved and heartened to know that today we start the process I have sworn to defend for all of my adult life. For the past week we have been groping at straws—asking ourselves the agonizing questions as to how this could happen. For all of us, it has been a week full of haunting speculation.

Today, we start the process which was envisioned by our founding fathers—today we start an orderly due process designed to provide the citizens of this great land with accountability.

To insure that this process remains fully intact, upon my return from Beirut I urgently requested that the Secretary of Defense conduct an inquiry into events leading up to the terrorist act which took the precious lives of young Americans at 0622 on 23 October. We owe this to the loved ones of those who have been killed, to the American people, to the Congress of the United States—and, of tremendous importance to me—to our Marines—past, present and future.

With that said—first, Mr. Chairman, let me set the scene.

Our Marines are situated in the middle of Beirut International Airport—this is a highly active commercial airport—the international terminal for a country of over three million people. We are there as guests, not invaders, so our facilities are provided in coordination with the Lebanese government.

Picture, if you will, the commercial activity at this airport—people, cars, trucks, major new construction, repair, new drainage systems under construction. This is a civilian environment, a hub-bub of activity by civilians, not military.

It is not a tactical strong point as some may envision.

Our mission is not, in a direct sense, the physical security of the airport—that specific mission is assigned to the Lebanese Armed Forces. Our basic mission is presence, and the logical question is—how do you define presence. Well, first let me tell you that presence as a mission is not in any military dictionary. It is not a classic military mission.

But the chain of command at the time correctly took presence to mean—be visible—provide a backdrop of U.S. presence which would be conducive to the stability of Lebanon—a sovereign Nation with a duly constituted government. I guess the best description is that we are a visible manifestation of U.S. strength and resolve to Lebanon and to the free world.

Besides, given the area we occupied—the threat as described by all available intelligence sources—a highly active commercial environment which was literally crawling with civilians—it would have been impossible for the Commander to establish a hard point defense in a classic tactical sense.

Again, please remember—we were guests of a friendly Nation—not on occupation duty!

My remarks today will take into account my trip to Beirut immediately after the terrorist attack. They include background information on the mission of the Marines since their second entry into Beirut on 29 September 1982. This background is essential to a complete understanding of what happened and how it could have happened.

These remarks will avoid discussion of the political or diplomatic considerations of our presence in Lebanon. It is not the place of a Marine to discuss those imperatives for military employment.

My remarks are based upon historical facts as I know them at this time.

The facts clearly show that our presence in Lebanon has gone through phases, each different, but clearly identifiable in the kaleidoscope of events over the past year. Each phase saw the threat to the security of the Marines ebb and flow in form and scope. Because of ever-changing circumstances and events, our forces banked heavily on the information, and warnings of danger, from the intelligence community—national and multinational. On the basis of this information, the Marines sought to anticipate events and take protective measures rather than just react after an event.

Subsequent to the successful evacuation of the PLO from Beirut, from 25 August to 9 September, the Marines withdrew from Lebanon on 10 September 1982, and resumed their normal duties as part of the Sixth Fleet in the Mediterranean.

The Marines (as part of the MNF) returned to Lebanon on 29 September 1982. The government of Lebanon requested the MNF to restore order after the assassination of their President, Bashir Gemeyal, and the tragic massacre of Palestinian refugees at the Sabra and Shatila camps. The Lebanese Armed Forces alone were clearly unable to exercise control in Beirut and the surrounding countryside.

As I mentioned previously, the assigned mission of the MNF, simply stated, was "presence."

It should be clearly understood that this was basically a diplomatic/political mission, not a military one in the classic sense, and the positioning of Marine forces at Beirut International Airport was not driven by tactical considerations. Moreover, the threats at the time, as reported to the Marines by the intelligence available did not require tactical deployment. Indeed, the mission of "presence" mitigated against such measures. Put another way, the Marines had to be seen by the Lebanese people.

The major threat when Marines returned to Beirut was from the accidental detonation of over 100,000 pieces of unexploded ordnance. This ordnance lay strewn and buried in and around the airfield, which was the "no-man's-land" during the battle of Beirut.

The rules of engagement under which the Marines were to operate were carefully constructed and promulgated by the Operational Commander. These were normal peacetime rules of engagement. They were restrictive in nature, but provided the right of self-protection and self-defense. The mission and rules of engagement were considered appropriate and adequate for the environment and threat.

During this phase the Marines were warmly greeted by the Lebanese people. Ordnance clearing operations and civic action projects undertaken by the Marines were appreciated by the populace. The overall security of BIA was, however, and still is, the responsibility of the LAF. In compliance with the mission, Marine dispositions were made at the airport to accommodate the LAF and to facilitate construction projects which would return the airport to normal use. Until November all went well.

By the beginning of November we entered a new and more ominous phase. The Marines had received intelligence reports that the passive threat environment of the previous month had changed. The intelligence community reported the problems from dissidents had become not just a possibility, but a probability. Armed with this intelligence, the Commander on the scene prudently initiated a variety of defensive measures.

The combination of warning and preparation paid off on 1 November 1982, when a 300-pound car-bomb was exploded on the main thoroughfare near the beach area, over which Marines received supplies from the ships offshore. A review of the measures previously taken to safeguard the beach area reveals that the Commander had:

- Hardened beach positions.
- Dispersed the beach support facilities.
- Limited access to the beach.
- Segregated all Marines from civilians (vendors, autos, etc.).
- Increased alert.
- Provided mobility at each checkpoint.

The terrorist effort was clumsy, amateurish and a failure. With the failure of the car-bomb, no further incidents occurred against the Marines during this period.

It was also during this phase that the Marines were authorized to begin an informal program to assist in training the LAF. This help was part of an overall effort to create a viable military entity which could eventually assume the security responsibilities for Beirut and later expand to greater Lebanon. While in itself this training effort may not be germane to the October bombing incident, it may be relative to the local perception of the role of the Marines in Beirut. Some may have perceived that Americans were no longer exclusively in a "presence" role; that we were in an assistance role. Motorized patrols were also initiated during this phase, and were conducted in east Beirut to provide clear visibility of the American presence.

In the latter part of December 1982, Marines were ordered not to patrol the Old Sidon road southwest of the airport. The rationale for this restriction is not immediately relative to the issue at hand, but it should be remembered that by the first of the year, for all practical purposes, the Marine perimeter was limited to the commercial complex and runways of BIA.

In February 1983, the low-threat environment continued. The Marines participated in a major rescue operation during a severe blizzard in the mountains east of Beirut. As part of the MNF, they operated beyond Syrian lines, rescuing Christians and Moslems. These operations helped sustain the principle of evenhandedness and confessional neutrality. It also had a major impact on the credibility of our "presence."

A new phase of the Beirut story began in March of this year. The situation began to deteriorate somewhat at the time. Although there was no intelligence of an increased threat, a 12-man Marine presence patrol in a community north of the airfield was attacked by a grenade thrown from a building. It resulted in five Marines receiving minor wounds. It was not predicted and was the first such incident against the Marines in four-

and-a-half months. Commanders immediately increased passive defense measures such as varying patrol routes, times, and size of patrols. In the opinion of some, about this time several Lebanese factions may have perceived a subtle shift of the USMNF from being pro-Lebanese to pro-Christian.

In April, the tragic car-bombing of the U.S. Embassy took place. Clearly, the United States was emerging as a prime target for those who either opposed or misinterpreted the role of the MNF in Lebanon. The former reason is more likely, in that the Italians and the French were also victims of terrorist harassment, even though they were not significantly involved in the reconstitution of the LAF.

Although I know of no intelligence warning which indicated that the spectacular car-bombing of the Embassy was in the offing, there were renewed warnings that terrorist attacks were likely to continue. Car-bombs were viewed as a likely form of attack. The Embassy bombing prompted several decisive steps to counter the threat. The Marines provided a special protective detail for the temporary U.S. Embassy and initiated a significant number of increased security measures.

• A Marine unit provided security and constructed defensive obstacles at the Ambassador's residence.

• Intensive efforts were made to increase intelligence sources in the immediate area of the Marine positions.

As pertains to the headquarters area:

• Civilian traffic was prohibited from the headquarters area.

• All civilian and military vehicles were searched in the proximity of the headquarters complex.

• Headquarters buildings and facilities were sandbagged. Over a quarter million sandbags were emplaced.

• Barbed wire entanglements were emplaced throughout. Concertina wire was strung in the civilian parking lot in front of headquarters. While the civilian parking lot adjacent to the BLT Headquarters lot was ideally suited for a mine field in a tactical sense, mines were not emplaced as the lot served the commercial airport and was actively used by civilians.

• Additional Marines were posted in each guard post during high-threat periods.

• Roof-top surveillance was increased.

• Foot patrols were increased within the headquarters area.

• A metal sewer pipe barrier was placed in front of the BLT headquarters building.

• Tunnels beneath the headquarters complex were sealed.

• The use of night observation devices was increased.

• Magazines were inserted in weapons at appropriate posts. (This in addition to previous orders which directed Marines on the perimeter and on patrol to stand duty with loaded weapons.)

• Tank ditches were contemplated, but considered unnecessary and impractical in view of the threat existing at the time and the commercial nature of the airport.

Throughout May, Marines operated in a high-threat environment and continued to actively patrol, train the LAF, and improve security. During June, there was a noticeable deterioration in relations with some factions of the local population. This was exacerbated by the reinfiltration of PLO elements into neighborhoods surrounding the airport. Among other indications, verbal harassment was directed against Marine patrols. The first rocket and mortar attacks against the LAF in the BIA complex occurred. Spillover of stray rounds came into the Marine positions. The Marines fully recognized the increased threat posed by this firing, and continued to harden positions by emplacing sandbags and digging-in deeper. Marine and LAF patrols were also integrated. Intelligence now indicated that rockets and mortars were to continue to be a primary concern to the safety of the troops.

During August, the periodic rocket attacks did increase against LAF targets, with a continued spillover into Marine positions. It was decided at this time to move the remainder of the BLT support personnel, and reaction platoon (approximately 150 men), into the BLT headquarters building to afford maximum protection against small arms, mortar, rocket and artillery fires.

It should be pointed out that the building was chosen because during the earlier fighting for Beirut it endured furious Israeli artillery barrages without being destroyed. An earth tremor in June also failed to cause any structural damage. It should be also noted that in a 13-month period, no Marine billeted in the building was killed or injured due to incoming artillery, mortar, rockets or small arms.

In late August, armed conflict between the LAF and AMAL militia in West Beirut began in earnest. On 4 September the Israeli Defense Forces withdrew to the Awwali River, bringing active fighting and factional conflict to the Alayh and Chouf regions above Beirut. Sustained hostile fire, some directed primarily at our Marines, impacted at the airport with increasing frequency. The shooting in and around Beirut was at ammunition levels rivaling major battles of World War II—over a million artillery rounds. Our Marines took appropriate measures to harden their positions, increase their alert status, and to move all support personnel in the terminal complex into reinforced buildings for protection against this intense shelling. The shelling was sufficient to halt all operations at the airport. When fired upon, the USMNF fired at specific targets with appropriate counterbattery fire.

Our naval gunfire support for the LAF was a major influence on the subsequent ceasefire, but unfortunately some could conclude that it may have increased the Moslem perception that our Marines were pro-Christian and no longer neutral. Be that as it may, I am in no position to judge. During this trying period of heavy fighting, the intelligence community continued to carry terrorist attacks as an active threat, but the threat was nonspecific and general, and overshadowed by the very specific and active reality of conventional military action.

The ceasefire on 26 September brought a fragile and uneasy peace to Beirut, but sniping at Marines became a daily occurrence. This brought us to a new phase. The warnings of the terrorist threat resurfaced from the intelligence community. While terrorist bomb intelligence continued to be non-precise, the focus of attention appeared to be the threat of car bombs to convoys providing support to the US diplomatic community in Beirut. Roughly 100 car bomb possibilities were developed since 1 June 1983. In some instances, the makes, colors and license numbers were provided. Marines on security duty received this information. All U.S., French, British and LAF units were looking for suspicious automobiles—

particularly as they related to convoys. The threat became a reality on 19 October 1983, when a car bomb was detonated in an attempt to impede a Marine supply convoy enroute to the temporary Embassy about 12km from the headquarters. Alertness and protective measures already taken by the Marines minimized the results of this attack. Like the car bomb at the beach the previous November, the attack against the convoy can be judged a failure.

At the same time, it must also be remembered that the cease fire was beginning to break down. Artillery fire in the Chouf was intermittently resumed, as was the small arms fire against the Marines. The terrorist threat remained vague while the active threat from artillery and small arms was increasing.

What I have been attempting to paint for you is a picture of Beirut for the past year. It is only within the framework of that picture can you have any hope of understanding the tragic events of a week ago.

A world where violence and normalcy live side by side. Marines on a diplomatic mission—located in a busy airport complex whose traffic and congestion rival that of any city—a Marine unit whose well-being depends upon the intelligence furnished to protect itself. A Marine unit which for months has undergone the indignity of attack with discipline and forbearance. A force of men whose morale remains high in spite of all that has happened.

I would now like to describe what occurred on Sunday morning, October 23, and why we believe that only extraordinary security could have met that massive and unanticipated threat.

At daybreak, a five-ton capacity Mercedes truck (roughly the size of a large dump truck and a type commonly seen at the Beirut International Airport) entered a public parking lot adjacent to the four-story, steel-reinforced concrete and sandbagged building which housed the headquarters elements of BLT 1/8. After making a complete circle of the parking lot for acceleration, and while travelling at a high speed, this truck:

- crashed through the outer defense of a barbed wire emplacement,
- moved at high speed between two sandbagged sentry posts,
- passed through a gate in an iron fence—jumped over a sewer pipe which had been placed as an obstacle to impede the forward movement of vehicles,
- plowed through a sandbag barrier,
- hit with precision a four-foot wide passenger entry into the lobby where its cargo, estimated by the Defense Intelligence Agency to be 5,000 pounds of explosives, detonated.

The entire event, which can best be described as the delivery by a suicidal driver of a 5,000-pound truck-bomb at very high speed, took approximately six seconds from start to finish. Rough calculations indicate that it would require a massive concrete wall to stop a vehicle of this weight and travelling at this speed. It is of particular importance to note that the Commander's security was oriented toward the threat of the past several months, i.e., artillery, rockets, mortars, small arms and car bombs. In this context, his security efforts had been successful. Obviously, the Commander's security arrangements were inadequate to counter this form of "kamikaze" attack. But, we have yet to find any shred of intelligence which would have alerted a reasonable and prudent commander to this new and unique threat. There was not even the indication of a capability to undertake such a monumental and precise action. General Tannous, the Commander of the Lebanese Armed Forces, informed me that he cannot recall, in his vast experience, a terrorist attack of the type which hit the headquarters of BLT 1/8 on 23 October 1983. In his opinion, it represents a new and unique terrorist threat, one which could not have been reasonably anticipated by any Commander.

Almost simultaneously, a smaller vehicle approached an eight-story apartment building to the north of Beirut International Airport which housed the French contingent. Since this building is on a busy thoroughfare, there would be no reason to suspect its intention. As it approached the building, it accelerated, took a sharp right into the driveway, and forced entry into an underground garage—where it exploded. During a personal conversation, General Cann, the Commander of the French contingent of the MNF, informed me he had no intelligence which would have warned him of this threat, as did General Angioni, the Commander of the Italian contingent.

I believe it important to recognize that there is sufficient evidence to conclude that both incidents were not suicidal acts by some individual fanatic. They were instead, well planned and professionally executed acts of terrorism which appear designed to drive our U.S. presence from Lebanon.

To bring you up-to-date, it is my professional estimate that our Marines have been targeted for terrorism by highly professional non-Lebanese elements. In my view, these acts of violence will continue, and the perpetrators will carefully examine and analyze our vulnerabilities and make every effort to exploit them. In short, I firmly believe that highly sophisticated and well-trained terrorists will target our Marines in the months to come. Therefore, I do not believe that we can ever create an effective passive capability which can counter all forms of terrorism in Lebanon or anywhere else.

With the foregoing said, I will now discuss initiatives which are underway or contemplated for increased security. The 24th MAU is in the process of decreasing vulnerability associated with large concentrations of Marines. Specifically:

- LAF APC's have been positioned at the BIA terminal and at the traffic circle in front of the airport.
- Vehicle access to command posts is now restricted to emergency and military vehicles.
- Civilian pedestrian access to the command post has been restricted to one location.
- MAU units have been placed in an indefinite "Condition I" (highest level) alert status.
- All entrances to the command posts have been blocked and reinforced.
- All rifle companies have reinforced the perimeter fence lines adjacent to their positions.
- An additional .50 caliber machine gun has been positioned to cover avenues of approach into the command post.
- Additional guard posts have been established throughout the MAU area, and an additional rifle company was sent from Camp Lejeune to provide security during the period of the recovery operations.
- Mobile reconnaissance patrols with anti-tank weapons have been established within the BIA perimeter.

• BLT 2/6 command element arrived at Beirut on 25 October to replace BLT 1/8 command elements.

• Definitive action is underway to strengthen the 24th MAU positions and to reduce vulnerability to terrorist attacks by isolating and barricading command and control and support areas.

In summary, I believe that:

• Our security measures were not adequate to stop a large, heavily laden truck, loaded with 5,000 pounds of high explosive, travelling at a high speed and driven by a suicide driver, which executed the attack in seconds from start to finish. This "flying truck bomb" was an unprecedented escalation in the previous terrorist threat, both in size of the weapon and method of delivery. I must continue to emphasize, however, that under our current disposition, restrictions, and mission, we will always have vulnerabilities, and that the other side will make every effort to exploit them.

That completes my statement. I would like to make two final comments:

I recognize that there remain many unanswered questions and a great deal of confusion surrounding this tragic event. For example, it is reported that the Commander, Colonel Geraghty, stated that he received a warning of the threat two days before the incident. The following message from him clarifies what he actually said, and I quote: "Sir, comment made to media was a general statement on car bomb warnings. At the weekly intelligence meeting between MNF Intel Officers and the Office of Beirut Security (*Surete Generale*), a listing of suspected car bombs, complete with car descriptions and license plate numbers is disseminated to the MNF by security officials. These car descriptions are copied and disseminated to our posts. Since our arrival, at least 100 potential car bombs have been identified to the MNF. After the attack on our convoy on 19 October 1983, the car bomb threat was quite obviously real to the USMNF; however, specific information on how car bomb attacks were to be conducted (i.e., kamikaze) or a description of the large truck that conducted the attack on the BLT were never received by 24 MAU."

Another example is that I was reported to have stated last week that security was adequate. Here let me set the scene and the context in which my remarks were made.

Five thousand pounds of high explosives destroyed a four-story steel reinforced concrete building. It was a heap of rubble. For over 50 hours, day and night, young Marines clawed at steel and concrete—more to save the injured who were trapped at the time than to recover the dead. The emotional scars were already deep—"Why me?" they asked. "Why am I alive and my buddies are dead?"

Their Commandant was asked, "Was security adequate?" I replied yes—it was adequate to meet what any reasonable and prudent commander should have expected prior to dawn on Sunday, October 23, 1983. And, I want you to know in that atmosphere my remarks were directed to weary and frustrated Marines.

Let me phrase what I was saying in a different way:

If you were to ask me whether the security around the headquarters building was adequate to protect the occupants against a five-ton Mercedes truck carrying 5,000 pounds of explosives at high speed—my answer would be NO!

And, if you would ask me whether the Commander should have known, given the explosion in the Embassy in April, my answer again would be NO! Both instances involved a terrorist bombing from a motor vehicle, but there the similarity ends. The delivery system was totally different as was every other aspect of the two incidents.

For these reasons, Mr. Chairman, I urgently requested the inquiry previously mentioned to determine the facts in an atmosphere that is conducive to such an inquiry. Knowing the Secretary of Defense as I do, and the respect I have for Admiral Long, there is no question in my mind that it will be a complete and thorough examination of this awful tragedy. I suggest we all await the board's findings.

I could not conclude my report to you without addressing the manner in which we reported our casualties. I know of your concern and share it. Our procedures have appeared to be excruciatingly slow. Please understand that in the impact of the destruction of the BLT Headquarters, and the tragic loss of life, our casualty reporting procedures for BLT 1/8 were destroyed. The requirements placed on the survivors to extricate and evacuate killed and wounded Marines as soon as possible, and the necessity to proceed slowly with regard to reporting for the sake of accurate identification and notification of the next of kin, were staggering. Due to the size of the task at hand and the painfully slow progress in this regard, the decision was made to release the names of those Marines who survived this disaster. We did not do this before for obvious reasons. The process was slow, mainly because of the need for complete accuracy. We didn't want to hurt anyone needlessly. Marines and members of your staffs worked tirelessly to ensure that timely and accurate information was released. The enormity of the situation is still upon us, and no one could feel more remorse than I over the prolonged suffering caused to many families by unavoidable delays in notifying them of their loved one's status.

The Marine Corps is proud of many things, but nothing more than the way we take care of our own. I want each of you to know that everything humanly possible is being done to facilitate the process. I would like to thank you and your staff for your assistance and understanding.

In closing, Mr. Chairman, let me say that the subject of increased terrorism against all Americans around the world may be one of the most serious problems which could be addressed by this Committee on a priority basis. This unprecedented, massive "kamikaze" attack was not against young Marines, sailors, and soldiers—it was a vicious, surprise attack against the United States of America and all we stand for in the free world.

Let me say, with all of the emphasis I can, that there are skilled and professional terrorists out there right now who are examining our vulnerabilities and making devices which are designed to kill Americans, lots of Americans around the world, in further acts of mass murder by terrorism. Let there be no doubt about it.

I would hope that the Congress would use this incident of cruel and premeditated mass murder to help us determine way which tell nations that they cannot export and support terrorists who kill innocent Americans with impunity.

The perpetrators and supporters of this challenge to the rights of free men everywhere must be identified and punished. I will have little sleep until this happens.

Thank you.

Appendix E

Long Commission Conclusions and Recommendations

All conclusions and recommendations of the Commission from each substantive part of this report are presented below.

Part One - The Military Mission

A. Mission Development and Execution

(1) Conclusion:

(a) The Commission concludes that the "presence" mission was not interpreted the same by all levels of the chain of command and that perceptual differences regarding that mission, including the responsibility of the USMNF for the security of Beirut International Airport, should have been recognized and corrected by the chain of command.

B. The Expanding Military Role

(1) Conclusion:

(a) The Commission concludes that U.S. decisions as regards Lebanon taken over the past fifteen months have been, to a large degree, characterized by an emphasis on military options and the expansion of the U.S. military role, nothwithstanding the fact that the conditions upon which the security of the USMNF were based continued to deteriorate as progress toward a diplomatic solution slowed. The Commission further concludes that these decisions may have been taken without clear recognition that these initial conditions had dramatically changed and that the expansion of our military involvement in Lebanon greatly increased the risk to, and adversely impacted upon the security of, the USMNF. The Commission therefore concludes that there is an urgent need for reassessment of alternative means to achieve U.S. objectives in Lebanon and at the same time reduce the risk to the USMNF.

(2) Recommendation:

(a) The Commission recommends that the Secretary of Defense continue to urge that the National Security Council undertake a reexamination of alternative means of achieving U.S. objectives in Lebanon, to include a comprehensive assessment of the military security options being developed by the chain of command and a more vigorous and demanding approach to pursuing diplomatic alternatives.

Part Two - Rules of Engagement (ROE)

ROE Implementation

(1) Conclusions:

(a) The Commission concludes that a single set of ROE providing specific guidance for countering the type of vehicular terrorist attacks that destroyed the U.S. Embassy on 18 April 1983 and the BLT Headquarters building on 23 October 1983 had not been provided to, nor implemented by, the Marine Amphibious Unit Commander.

(b) The Commission concludes that the mission statement, the original ROE, and the implementation in May 1983 of dual "Blue Card - White Card" ROE contributed to a mind-set that detracted from the readiness of the USMNF to respond to the terrorist threat which materialized on 23 October 1983.

Part Three - The Chain of Command

A. Exercise of Command Responsibility by the Chain of Command Prior to 23 October 1983.

(1) Conclusions:

(a) The Commission is fully aware that the entire chain of command was heavily involved in the planning for, and support of, the USMNF. The Commission concludes however, that USCinCEur, CinCUSNavEur, COMSixthFlt and CTF 61 did not initiate actions to ensure the security of the USMNF in light of the deteriorating political/military situation in Lebanon. The Commission found a lack of effective command supervision of the USMNF security posture prior to 23 October 1983.

(b) The Commission concludes that the failure of the operational chain of command to correct or amend the defensive posture of the USMNF constituted tacit approval of the security measures and procedures in force at the BLT headquarters building on 23 October 1983.

(c) The Commission further concludes that although it finds the USCinCEur operational chain of command at fault, it also finds that there was a series of circumstances beyond the control of these commands that influenced their judgement and their actions relating to the security of the USMNF.

Recommendation:

(a) The Commission recommends that the Secretary of Defense take whatever administrative or disciplinary action he deems appropriate, citing the failure of the USCinCEur operational chain of command to monitor and supervise effectively the security measures and procedures employed by the USMNF on 23 October 1983.

Part Four - Intelligence

A. Intelligence Support

(1) Conclusions:

(a) The Commission concludes that although the USMNF Commander received a large volume of intelligence warnings concerning potential terrorist threats prior to 23 October 1983, he was not provided with the timely intelligence, tailored to his specific operational needs, that was necessary to defend against the broad spectrum of threats he faced.

(b) The Commission further concludes that the HUMINT support to the USMNF Commander was ineffective, being neither precise nor tailored to his needs. The Commission believes that the paucity of U.S. controlled HUMINT provided to the USMNF Commander is in large part due to policy decisions which have resulted in a U.S. HUMINT capability commensurate with the resources and time that have been spent to acquire it.

(2) Recommendations:

(a) The Commission recommends that the Secretary of Defense

establish an all-source fusion center, which would tailor and focus all-source intelligence support to U.S. military commanders involved in military operations in areas of high threat, conflict or crisis.

(b) The Commission further recommends that the Secretary of Defense take steps to establish a joint CIA/DOD examination of policy and resource alternatives to immediately improve HUMINT support to the USMNF contingent in Lebanon and other areas of potential conflict which would involve U.S. military operating forces.

Part Five - Pre-Attack Security

A. Command Responsibility for the Security of the 24th MAU and BLT 1/8 Prior to 23 October 1983.

(1) Conclusions:

(a) The combination of a large volume of specific threat warnings that never materialized and the perceived and real pressure to accomplish a unique and difficult mission contributed significantly to the decision of the MAU and BLT Commanders regarding the security of their force. Nevertheless, the Commission concludes that the security measures in effect in the MAU compound were neither commensurate with the increasing level of threat confronting the USMNF nor sufficient to preclude catastrophic losses such as those that were suffered on the morning of 23 October 1983. The Commission further concludes that while it may have appeared to be an appropriate response to the indirect fire being received, the decision to billet approximately one quarter of the BLT in a single structure contributed to the catastrophic loss of life.

(b) The Commission concludes that the BLT Commander must take responsibility for the concentration of approximately 350 members of his command in the BLT Headquarters building, thereby providing a lucrative target for attack. Further, the BLT Commander modified prescribed alert procedures, thereby degrading security of the compound.

(c) The Commission also concludes that the MAU Commander shares the responsibility for the catastrophic losses in that he condoned the concentration of personnel in the BLT Headquarters building, concurred in the modification of prescribed alert procedures, and emphasized safety over security in directing that sentries on Posts 4, 5, 6, and 7 would not load their weapons.

(d) The Commission further concludes that although it finds the BLT and MAU Commanders to be at fault, it also finds that there was a series of circumstances beyond their control that influenced their judgement and their actions relating to the security of the USMNF.

(2) Recommendation:

(a) The Commission recommends that the Secretary of Defense take whatever administrative or disciplinary action he deems appropriate, citing the failure of the BLT and MAU Commanders to take the security measures necessary to preclude the catastrophic loss of life in the attack on 23 October 1983.

Part Seven - Post-Attack Security

Redeployment, Dispersal and Physical Barriers

(1) Conclusions:

(a) The Commission concludes that the security measures taken since 23 October 1983 have reduced the vulnerability of the USMNF to catastrophic losses. The Commission also concludes, however, that the security measures implemented or planned for implementation for the USMNF as of 30 November 1983, were not adequate to prevent continuing significant attrition of the force.

(b) The Commission recognizes that the current disposition of USMNF forces may, after careful examination, prove to be the best available option. The Commission concludes, however, that a comprehensive set of alternatives should be immediately prepared and presented to the National Security Council.

(2) Recommendation:

(a) Recognizing that the Secretary of Defense and the Joint Chiefs of Staff have been actively reassessing the increased vulnerability of the USMNF as the political/military environment in Lebanon has changed, the Commission recommends that the Secretary of Defense direct the operational chain of command to continue to develop alternative military options for accomplishing the mission of the USMNF while reducing the risk to the force.

Part Eight - Casualty Handling

A. On-Scene Medical Care

(1) Conclusion:

(a) The Commission concludes that the speed with which the on-scene U.S. military personnel reacted to rescue their comrades trapped in the devastated building and to render medical care was nothing short of heroic. The rapid response by Italian and Lebanese medical personnel was invaluable.

B. Aeromedical Evacuation/Casualty Distribution

(1) Conclusions:

(a) The Commission found no evidence that any of the wounded died or received improper medical care as a result of the evacuation or casualty distribution procedures. Nevertheless, the Commission concludes that overall medical support planning in the European theater was deficient and that there was an insufficient number of experienced medical planning staff officers in the USCinCEur chain of command.

(b) The Commission found that the evacuation of the seriously wounded to U.S. hospitals in Germany, a transit of more than four hours, rather than to the British hospital in Akrotiri, Cyprus, a transit of one hour, appears to have increased the risk of those patients. Similarly, the Commission found that the subsequent decision to land the aircraft at Rhein Main rather than Ramstein, Germany, may have increased the risk to the most seriously wounded. In both instances, however, the Commission has no evidence that there was an adverse medical impact on the patients.

(2) Recommendations:

(a) The Commission recommends that the Secretary of Defense direct the Joints Chiefs of Staff, in coordination with the Services, to review medical plans and staffing of each echelon of the operational and administrative chains of command to ensure appropriate and adequate medical support of the USMNF.

(b) The Commission further recommends that the Secretary of Defense direct USCinCEur to conduct an investigation of the decisions made regarding the destination of aeromedical evacuation aircraft and the distribution of casualties on 23 October 1983.

C. Definitive Medical Care

(1) Conclusion:

(a) The Commission concludes that the definitive medical care provided the wounded at the various treatment facilities was excellent, and that as of 30 November 1983, there is no evidence of any mortality or morbidity resulting from inappropriate or insufficient medical care.

D. Israeli Offer of Medical Assistance

(1) Conclusion:

(a) The Commission found no evidence that any factor other than the desire to provide immediate, professional treatment for the wounded influenced decisions regarding the Israeli offer; all offers of assistance by Israel were promptly and properly referred to the theater and on-scene commanders. At the time the initial Israeli offer was reviewed by CTF 61, it was deemed not necessary because the medical capabilities organic to CTF 61 were operational and functioning adequately, the RAF hospital at Akrotiri was mobilized and ready, and sufficient U.S. and RAF medical evacuation aircraft were enroute.

E. Identification of the Dead

(1) Conclusion:

(a) The Commission concludes that the process for identification of the dead following the 23 October 1983 catastrophe was conducted very efficiently and professionally, despite the complications caused by the destruction and/or absence of identification data.

(2) Recommendation:

(a) The Commission recommends that the Secretary of Defense direct the creation of duplicate medical/dental records, and assure availability of fingerprint files, for all military personnel. The Commission further recommends that the Secretary of Defense direct the Service Secretaries to develop jointly improved, state-of-the-art identification tags for all military personnel.

Part Nine - Military Response to Terrorism

A. A Terrorist Act

(1) Conclusion:

(a) The Commission concludes that the 23 October 1983 bombing of the BLT Headquarters building was a terrorist act sponsored by sovereign States or organized political entities for the purpose of defeating U.S. objectives in Lebanon.

B. International Terrorism

(1) Conclusion:

(a) The Commission concludes that international terrorist acts endemic to the Middle East are indicative of an alarming world-wide phenomenon that poses an increasing threat to U.S. personnel and facilities.

C. Terrorism as a Mode of Warfare

(1) Conclusion:

(a) The Commission concludes that state sponsored terrorism is an important part of the spectrum of warfare and that adequate response to this increasing threat requires an active national policy which seeks to deter attack or reduce its effectiveness. The Commission further concludes that this policy needs to be supported by political and diplomatic actions and by a wide range of timely military response capabilities.

(2) Recommendation:

(a) The Commission recommends that the Secretary of Defense direct the Joint Chiefs of Staff to develop a broad range of appropriate military responses to terrorism for review, along with political and diplomatic actions, by the National Security Council.

D. Military Preparedness

(1) Conclusion:

(a) The Commission concludes that the USMNF was not trained, organized, staffed, or supported to deal effectively with the terrorist threat in Lebanon. The Commission further concludes that much needs to be done to prepare U.S. military forces to defend against and counter terrorism.

(2) Recommendation:

(a) The Commission recommends that the Secretary of Defense direct the development of doctrine, planning, organization, force structure, education and training necessary to defend against and counter terrorism.

Appendix F
American Deaths in Beirut

Corporal Terry W. Abbott
Lance Corporal Clemon Alexander
Private First Class John R. Allman
Corporal Moses Arnold, Jr.
Private First Class Charles K. Bailey

Lance Corporal Nicholas Baker
Lance Corporal Johansen Banks
Lance Corporal Richard E. Barrett
Hospital Corpsman First Class Ronny K. Bates, USN
First Sergeant David L. Battle

Lance Corporal James R. Baynard
Hospitalman Jesse W. Beamon, USN
Gunnery Sergeant Alvin Belmer
Private First Class Shannon D. Biddle
Private First Class Stephen B. Bland

Corporal Richard L. Blankenship
Private First Class John W. Blocker
Captain Joseph J. Boccia, Jr.
Corporal Leon W. Bohannon
Staff Sergeant John R. Bohnet, Jr.

Corporal John J. Bonk, Jr.
Lance Corporal Jeffrey J. Boulos
Corporal David R. Bousum
First Lieutenant John N. Boyett
Corporal Anthony K. Brown

Lance Corporal David W. Brown
Lance Corporal Bobby B. Buchanan, Jr.
Corporal John B. Buckmaster
Private First Class William F. Burley
Major Alfred L. Butler III

Hospitalman Jimmy R. Cain, USN
Corporal Paul L. Callahan
Corporal Mecot E. Camara
Private First Class Bradley J. Campus
Major Randall A. Carlson, USA

Lance Corporal Johnnie D. Ceasar
Lance Corporal Sam Cherman
Lance Corporal Randy W. Clark
Private First Class Marc L. Cole
Specialist Four Marcus E. Coleman, USA

Private First Class Juan M. Comas
Sergeant Robert A. Conley
Corporal Charles D. Cook
Lance Corporal Curtis J. Cooper
Lance Corporal Johnny L. Copeland

Corporal Bert D. Corcoran
Lance Corporal David L. Cosner
Sergeant Kevin P. Coulman
Sergeant Manuel A. Cox
Lance Corporal Brett A. Croft

Lance Corporal Rich R. Crudale
Lance Corporal Kevin P. Custard
Lance Corporal Russell E. Cyzick
Corporal David L. Daugherty
Major Andrew L. Davis

Private First Class Sidney J. Decker
Private First Class Michael J. Devlin
Corporal Thomas A. Dibenedetto
Private First Class Nathaniel G. Dorsey
Sergeant Major Frederick B. Douglass

Lance Corporal George L. Dramis
Corporal Timothy J. Duinnigan
Hospitalman Bryan L. Earle, USN
Master Sergeant Roy L. Edwards
Hospital Corpsman Third Class William D. Elliott, Jr., USN

Lance Corporal Jesse J. Ellison
Private First Class Danny R. Estes
Private First Class Sean F. Estler
Lance Corporal Thomas A. Evans
Hospital Corpsman Third Class James E. Faulk, USN

Private First Class Richard A. Fluegel
Corporal Steven M. Forrester
Hospital Corpsman Third Class William B. Foster, Jr., USN
Corporal Michael D. Fulcher
Lance Corporal Benjamin E. Fuller

Lance Corporal Michael S. Fulton
Corporal William R. Gaines, Jr.
Lance Corporal Sean R. Gallagher
Lance Corporal David B. Gander
Lance Corporal George M. Gangur

Staff Sergeant Leland E. Gann
Lance Corporal Randall J. Garcia
Staff Sergeant Ronald J. Garcia
Sergeant Edward J. Gargano
Lance Corporal David D. Gay

Staff Sergeant Harold D. Ghumm
Lance Corporal Warner Gibbs, Jr.
Corporal Timothy R. Giblin
Chief Electronics Technician Michael W. Gorchinski, USN
Lance Corporal Richard J. Gordon

Lance Corporal Harold F. Gratton
Sergeant Robert B. Greaser
Lance Corporal Davin M. Green
Lance Corporal Thomas A. Hairston
Sergeant Freddie L. Haltiwanger, Jr.

Lance Corporal Virgel D. Hamilton
Sergeant Gilbert Hanton
Lance Corporal William Hart
Captain Michael S. Haskell
Private First Class Michael A. Hastings

Lance Corporal Jeffrey T. Hattaway
Captain Paul A. Hein
Lance Corporal Douglass E. Held
Private First Class Mark A. Helms
Lance Corporal Ferrandy D. Henderson

Gunnery Sergeant Matilde Hernandez, Jr.
Lance Corporal Rodolfo Hernandez
Corporal Stanley G. Hester
Gunnery Sergeant Donald W. Hildreth
Staff Sergeant Richard H. Holberton

Hospital Corpsman Third Class Robert S. Holland, USN
Lance Corporal Bruce A. Hollingshead
Private First Class Melvin D. Holmes
Corporal Bruce L. Howard
Lieutenant John R. Hudson, USNR

Corporal Terry L. Hudson
Lance Corporal Lyndon J. Hue
Second Lieutenant Maurice E. Hukill
Lance Corporal Edward S. Iacovino, Jr.
Private First Class John J. Ingalls

Warrant Officer Paul G. Innocenzi III
Lance Corporal James J. Jackowski
Lance Corporal Jeffrey W. James
Lance Corporal Nathaniel W. Jenkins
Hospital Corpsman Second Class Michael H. Johnson, USN

Corporal Edward A. Johnston
Lance Corporal Steven Jones
Private First Class Thomas A. Julian
Hospital Corpsman Second Class Marion E. Kees, USN
Sergeant Thomas C. Keown

Gunnery Sergeant Edward E. Kimm
Lance Corporal Walter V. Kingsley
Sergeant Daniel S. Kluck, USA
Lance Corporal James C. Knipple
Lance Corporal Todd A. Kraft

Lance Corporal Freas H. Kreischer III
Lance Corporal Keith J. Laise
Lance Corporal Thomas G. Lamb
Lieutenant Mark A. Lange, USN
Lance Corporal James J. Langon IV

Sergeant Michael S. Lariviere
Corporal Stephen B. Lariviere
Master Sergeant Richard L. Lemnah
Corporal David A. Lewis
Sergeant Val S. Lewis

Corporal Joseph R. Livingston
Second Lieutenant Donald G. Losey, Jr.
Lance Corporal Paul D. Lyon, Jr.
Major John W. Macroglou
Corporal Samuel Maitland

Staff Sergeant Charlie R. Martin
Private First Class Jack L. Martin
Corporal David S. Massa
Corporal Michael R. Massman
Private Joseph J. Mattacchione

Staff Sergeant Ben Henry Maxwell, USA
Lance Corporal John McCall
Corporal James E. McDonough
Private First Class Timothy R. McMahon
Corporal Robert V. McMaugh

Lance Corporal Timothy D. McNeely
Hospital Corpsman Second Class George N. McVicker II, USN
Private First Class Louis Melendez
Corporal Richard H. Menkins II
Corporal Michael D. Mercer

Lance Corporal Ronald W. Meurer
Hospital Corpsman Third Class Joseph P. Milano, USN
Corporal Joseph P. Moore
Lance Corporal Richard A. Morrow
Lance Corporal John F. Muffler

Private First Class Alex Munoz
Corporal Harry D. Myers
First Lieutenant David J. Nairn
Lance Corporal Luis A. Nava
Captain Michael J. Ohler

Corporal John A. Olson
Private First Class Robert P. Olson
Staff Sergeant Alexander M. Ortega
Chief Warrant Officer Richard C. Ortiz
Private First Class Jeffrey B. Owen

Corporal Joseph A. Owens
Corporal Ray Page
Lance Corporal Ulysses G. Parker
Lance Corporal Mark W. Payne
Gunnery Sergeant John L. Pearson

Lance Corporal Marvin H. Perkins
Private First Class Thomas S. Perron
Sergeant John A. Phillips, Jr.
Chief Hospital Corpsman George W. Piercy, USN
First Lieutenant C. Wayne Plymel

Sergeant William H. Pollard
Sergeant Rafael Pomalestorres
Corporal Victor M. Prevatt
Private First Class James C. Price
Staff Sergeant Patrick K. Prindeville

Private First Class Eric C. Pulliam
Hospital Corpsman Third Class Diomedes J. Quirante, USN
Lance Corporal David M. Randolph
Gunnery Sergeant Charles R. Ray
Corporal David L. Reagan

Private First Class Rui A. Relvas
Private First Class Terrence L. Rich
Lance Corporal Warren Richardson
Sergeant Juan C. Rodriguez
Lance Corporal Louis J. Rotondo

Staff Sergeant Mark E. Salazar, USA
Lance Corporal Guillermo San Pedro, Jr.
Lance Corporal Michael C. Sauls
First Lieutenant Charles J. Schnorf
Private First Class Scott L. Schultz

Captain Peter J. Scialabba
Corporal Gary R. Scott
Corporal Ronald L. Shallo
Lance Corporal Thomas A. Shipp
Private First Class Jerryl D. Shropshire

Lance Corporal James F. Silvia
Lance Corporal Stanley J. Sliwimski
Lance Corporal Kirk H. Smith
Staff Sergeant Thomas G. Smith
Captain Vincent L. Smith

Lance Corporal Edward Soares
Sergeant Allen H. Soifert
First Lieutenant William S. Sommerhof
Lance Corporal Michael C. Spaulding
Lance Corporal John W. Spearing

Lance Corporal Stephen E. Spencer
Private First Class Bill J. Stelpflug
Private First Class Horace R. Stephens
Private First Class Craig S. Stockton
Lance Corporal Jeffrey G. Stokes

Lance Corporal Thomas D. Stowe
Lance Corporal Eric D. Sturghill
Lance Corporal Devon L. Sundar
Lieutenant James F. Surch, USN
Corporal Dennis A. Thompson

Staff Sergeant Thomas P. Thorstad
Private First Class Stephen D. Tingley
Lance Corporal John J. Tishmack
Corporal Henry Townsend, Jr.
Private Lex D. Trahan

Master Sergeant Richard Twine, USA
Corporal Pedro J. Valle
Private First Class Donald D. Vallone, Jr.
Intelligence Specialist First Class Michael R. Wagner, USN
Lance Corporal Eric R. Walker

Lance Corporal Leonard W. Walker
Corporal Eric G. Washington
Corporal Obrian Weekes
Chief Warrant Officer Kenneth V. Welch, USA
First Sergeant Tandy W. Wells

Lance Corporal Steven B. Wentworth
Sergeant Allen D. Wesley
Gunnery Sergeant Lloyd D. West
Staff Sergeant John R. Weyl
Corporal Burton D. Wherland

Lance Corporal Dwayne W. Wigglesworth
Lance Corporal Rodney J. Williams
Gunnery Sergeant Scipio Williams, Jr.
Lance Corporal Johnny A. Williamson
Captain Walter A. Wint, Jr.

Captain William W. Winter
Corporal John E. Wolfe
First Lieutenant Donald E. Woollett
Hospital Corpsman Third Class David E. Worley, USN
Private First Class Craig L. Wyche

Private First Class James G. Yarber, USA
Corporal Jeffrey D. Young
First Lieutenant William A. Zimmerman

Appendix G
Unit Commendations

The Secretary of the Navy takes pleasure in presenting the NAVY UNIT COMMENDATION to

MEDITERRANEAN AMPHIBIOUS READY GROUP 2-82
AND
THIRTY-SECOND MARINE AMPHIBIOUS UNIT

for service as set forth in the following

CITATION:

For exceptionally meritorious service in a mission of great importance to the Government of the United States from 16 August 1982 to 10 September 1982 which resulted in the cessation of armed conflict between Israeli and Syrian and Palestine Liberation Organization (PLO) Forces. On 25 August 1982, elements of Mediterranean Amphibious Ready Group 2-82 and THIRTY-SECOND Marine Amphibious Unit conducted a flawless landing from the sea into the Port of Beirut, Lebanon. On occupying the Port, elements of the THIRTY-SECOND Marine Amphibious Unit, serving as the United States contingent of a multinational force, immediately commenced the supervision of the evacuation of 6,436 PLO and Syrian combatants. The superior performance and tireless dedication of all personnel reflected the epitome of professionalism and exceeded the performance normally expected. The total success of the mission contributed visibly and significantly to the United States' objectives of world peace and Middle East stability. By their steadfast courage, recourcefulness, and unwavering devotion to duty, the officers and enlisted personnel of Mediterranean Amphibious Ready Group 2-82 and THIRTY-SECOND Marine Amphibious Unit reflected great credit upon themselves and upheld the highest traditions of the Marine Corps and the United States Naval Service.

/s/ John Lehman
Secretary of the Navy

The Secretary of the Navy takes pleasure in presenting the NAVY UNIT COMMENDATION to

MEDITERRANEAN AMPHIBIOUS READY GROUP 3-82
AND
MARINE AMPHIBIOUS UNIT 24

for service as set forth in the following

CITATION:

For exceptionally meritorious service from 29 October 1982 to 15 February 1983 in a mission of great national and international importance while serving as the United States Forces Ashore Lebanon and supporting forces of the Multinational Force peace initiatives in Lebanon. On 29 October 1982, Mediterranean Amphibious Ready Group 3-82 and Marine Amphibious Unit 24 conducted a combined surface and helo landing across Black Beach and the International Airport of Beirut, Lebanon and immediately commenced coordinated motorized and foot patrols with other Multinational Force contingents throughout the City of Beirut. Simultaneously with operations ashore, Amphibious Task Force helicopters provided special support to the U.S. Ambassador to Lebanon and special Presidential envoys, and provided major logistics lifts from Air Support Head at Larnaca, Cyprus, to Beirut. Their superior performance and tireless devotion contributed significantly to the national objectives of world peace and Middle East stability. By their resolute determination, unrelenting perseverance, and steadfast dedication to duty, the officers and enlisted personnel of Mediterranean Amphibious Ready Group 3-82 and Marine Amphibious Unit 24 reflected great credit upon themselves and upheld the highest traditions of the Marine Corps and the United States Naval Service.

/s/ John Lehman
Secretary of the Navy

The Secretary of the Navy takes pleasure in presenting the NAVY UNIT COMMENDATION to

COMMANDER AMPHIBIOUS SQUADRON 2
AND
TWENTY-SECOND MARINE AMPHIBIOUS UNIT

for service as set forth in the following

CITATION:

For exceptionally meritorious service from 14 February 1983 to 30 May 1983 in a mission of great national and international importance while serving as the United States Forces Ashore Lebanon and supporting forces of the Multinational Peacekeeping Force in Lebanon. Shortly after arrival, the Amphibious Task Force was requested to conduct and carry out humanitarian relief operations to assist the Lebanese Government in dealing with the effects of a severe snowstorm in the mountainous areas of Central Lebanon. On 18 April 1983, following the catastrophic destruction of the U.S. Embassy in Beirut by terrorist bombing, Amphibious Task Force Units rapidly and decisively provided lifesaving assistance, security, and communications support and follow-on security for all Embassy operations in Beirut. Simultaneously with operations ashore, the Amphibious Task Force provided helicopter and special security support for the Secretary of State, U.S. Ambassador to Lebanon, and special Presidential Envoys. The total success of the mission contributed visibly and significantly to the national objectives of world peace and Middle East stability. By their resolute determination, steadfast perseverance, and selfless devotion to duty, the officers and enlisted personnel of Commander Amphibious Squadron TWO and TWENTY-SECOND Marine Amphibious Unit reflected great credit upon themselves and upheld the highest traditions of the Marine Corps and the United States Naval Service.

/s/ John Lehman
Secretary of the Navy

The Secretary of the Navy takes pleasure in presenting the NAVY UNIT COMMENDATION to

MEDITERRANEAN AMPHIBIOUS READY GROUP (MARG) 2-83
AND
24TH MARINE AMPHIBIOUS UNIT

for service as set forth in the following

CITATION:

For exceptionally meritorious service in support of U.S. peace initiatives in Lebanon from 28 May 1983 to 19 November 1983. Performing a difficult mission to ensure stability during a period of complex political and life-threatening conditions in Lebanon, the units of U.S. Peacekeeping Forces Lebanon displayed exceptional courage, resolve, and flexibility in providing supporting actions and evidence of U.S. national concern, often in the face of danger to personnel and equipment, to provide conditions in which the duly constituted Government of Lebanon could survive. Displaying superb dedication and flexibility in the face of hostile fire, the Navy and Marine Corps units ensured and supported conditions to permit the vital work of diplomacy to go forward and the Government of Lebanon to remain intact. With extraordinary heroism, the 24th Marine Amphibious Unit (24TH MAU) made possible conditions for national reconciliation in an area vital to U.S. national security. On 23 October 1983, the 24TH MAU suffered unprecedented personnel losses of approximately 330 killed and wounded as a result of terrorist bombing of the 24TH MAU Headquarters building. In the face of this adversity, they continued to fulfill their assigned mission while carrying out rescue and evacuation efforts. By their exceptional courage, perseverance, and steadfast devotion to duty, the officers and enlisted personnel of Mediterranean Amphibious Ready Group (MARG 2-83) and 24th Marine Amphibious Unit reflected great credit upon themselves and upheld the highest traditions of the Marine Corps and the United States Naval Service.

/s/ John Lehman
Secretary of the Navy

The Secretary of the Navy takes pleasure in presenting the NAVY UNIT COMMENDATION to

MEDITERRANEAN AMPHIBIOUS READY GROUP (MARG) 1-84
AND
22ND MARINE AMPHIBIOUS UNIT (MAU)

for service as set forth in the following

CITATION:

For exceptionally meritorious service against a heavily armed rebel force threatening the personal safety of American citizens and the established Government of Grenada, and in subsequent operations with the Multinational Force, Lebanon, from 20 October 1983 to 26 February 1983. Through calculated forethought and incisive action by the officers and men of Task Force ONE HUNDRED TWENTY-FOUR (Mediterranean Amphibious Ready Group [MARG] 1-84 and 22nd Marine Amphibious Unit [MAU]), the lives of hundreds of American civilians were saved, rebel forces were subdued, and the Government of Grenada restored. While serving with the Multinational Force, Lebanon, these units maintained a positive U.S. presence under the most demanding circumstances during a period of extremely dynamic, interrelated, and complex political instability. Demonstrating determined resolve in the face of open hostilities, MARG 1-84/22ND MAU dramatically improved defensive positions, provided support for diplomatic efforts, evacuated U.S. civilians, foreign nationals, and non-essential support personnel, and repositioned U.S. Multinational Forces. By their selfless determination, exceptional performance, personal sacrifice, and steadfast devotion to duty, the officers and enlisted personnel of Mediterranean Amphibious Ready Group (MARG) 1-84 and the 22nd Marine Amphibious Unit (MAU) reflected great credit upon themselves and upheld the highest traditions of the Marine Corps and the United States Naval Service.

/s/ John Lehman
Secretary of the Navy

The Secretary of the Navy takes pleasure in presenting the MERITORIOUS UNIT COMMENDATION to

TASK GROUP 61.8
AND
31ST MARINE AMPHIBIOUS UNIT

for service as set forth in the following

CITATION:

For meritorious service while serving as afloat support to U.S. Peacekeeping Forces Lebanon from 12 September 1983 to 10 October 1983 during their mission to ensure stability during a period of complex political and life-threatening conditions in Lebanon. Throughout this arduous period, Task Group 61.8 and 31st Marine Amphibious Unit maintained active presence offshore in a state of total readiness to reinforce when called upon for support. Their unrelenting commitment provided evidence of U.S. national concern to enable conditions in which the duly constituted government of Lebanon could survive. By their resolute determination, courage and complete dedication to duty, the officers and enlisted personnel of Task Force 61.8 and 31st Marine Amphibious Unit reflected credit upon themselves and upheld the highest traditions of the Marine Corps and the United States Naval Service.

/s/ John Lehman
Secretary of the Navy

Index

Key: **boldface type** = illustration; *n* = footnote; *ff* = following pages

Made in United States
Cleveland, OH
06 November 2026

47678356R10114